[= := − ÷ x + ∞]

ARE ALL BELIEVERS EQUAL?

THE POWER OF MORAL EXCELLENCE
[2 PETER 1:1-21]

HOLLIS L. GREEN

GlobalEdAdvance
Press

ARE ALL BELIEVERS EQUAL?

—The Power of Moral Excellence

Copyright © 2022 by Hollis L. Green

Library of Congress Control Number: 2022914745

ISBN 978-1-950839-11-7

Subject Codes and Description: 1: REL 00406100 Religion: Biblical Criticism & Interpretation –New Testament; 2. EDU 040000. Education: Philosophy, Theory and Social Aspects; 3: EDU 032000 Education: Mathematics / Applied.

All rights reserved, including the right to reproduce this book or any part thereof in any form, except for inclusion of brief quotations in a review, without the written permission of GlobalEdAdvance PRESS and the author. Old Testament scriptures not otherwise noted are from the NIV. New Testament scriptures are primarily from EDNT unless otherwise noted.

Cover by Global Graphics, NYC

Published Nashville, TN

Printed in Australia, Brazil, EU, France, Germany, Italy, Poland, Russia, Spain, UK (3 sites in USA) and available on the Espresso Book Machine© worldwide.

The Press does not have ownership of the contents of a book; this is the author's book, and the author owns the copyright. All theory, concepts, constructs, and perspectives are those of the author and not necessarily the Press. They are presented for open and free discussion of the issues involved. All comments and feedback should be directed to the Email: [comments4author@aol.com] and the comments will be forwarded to the author for response.

Order books from www.gea-books.com/bookstore/
or any place good books are sold.

Published by
GlobalEdAdvancePRESS
a division of
Global Educational Advance, Inc.
GlobalEdAdvance.org

DEDICATION

This work is dedicated with appreciation to the Scholarly Referees who read an early draft and made suggestions and offered encouragement:

Paratan Balloo, MBA, DPhil, Vice Chancellor/CEO, O.A.S.I.S. UNIVERSITY, Trinidad, W.I.

Ronald A. Berk, PhD, Professor Emeritus, Biostatistics & Measurement, and former Assistant Dean for Teaching, The Johns Hopkins University.

R. John Buuck, PhD, President Emeritus, Concordia University (Wisconsin)

Alfred S. Cockfield, DPhil, Senior Pastor, God's Battalion of Prayer Global NETWORK, Brooklyn, NY.

Coy Webb, MBA, ThD, Family Pastor, New Life Church, Bluff City, TN.

Are All Believers Equal?

The Power Of Moral Excellence

[2 Peter 1:1-21]

Contents

PART ONE: Front Material

Preface: 11
Hopeful Anticipation of Improved Communication

Introduction   23
Prior Learning Impacts Human Existence

Grounding: 39
Grafting into the Spiritual Rootstock

PART TWO: First Steps

Step I	Start with a Common Faith	53
Step II	Build on Precious Promises	69
Step III	Share Sincere Benevolent Love	79
Step IV	Multiply Brotherly Kindness	89

PART THREE: Intermediate Steps

Step V	Maintain Godly Worship	99
Step VI	Sustain Enduring Steadfastness	119
Step VII	Practice Self-control	127
Step VIII	Learn from Books and Teachers	135

PART FOUR: Final Steps

Step IX	Crown Faith with Moral Excellence	151
Step X	Advance a Kingdom Lifestyle	161
Step XI	Gain Kingdom Entrance	179
Step XII	Refresh your Memory of These Things	189

PART FIVE: Back Material

Afterword	Beware of Leapfrog Didactic Schemes (2 Peter 2:1-2 EDNT)	199

About the Author   209

APPENDIX A	Recent Books by the Author	211
APPENDIX B	Names and Attributes of God	216
APPENDIX C	How a Personal God Communicates	222
APPENDIX D	Faith-based Mentors and Coaches	224
APPENDIX E	Guidance for Converts	230
Bibliography and Reading List		232

PART ONE: Front Material

PREFACE
HOPEFUL ANTICIPATION OF IMPROVED COMMUNICATION (>)

INTRODUCTION
PRIOR LEARNING IMPACTS HUMAN EXISTENCE (^→)

GROUNDING
GRAFTING INTO THE SPIRITUAL ROOTSTOCK (<)

Are All Believers Equal?

The Power Of Moral Excellence

(>)
Preface:

Hopeful Anticipation of Improved Communication

(>) Greater or more than

The author has hopeful anticipation that Applied Mathematics and classical symbols used in this work will become a "value added" tool for use in the reader's written communication. Perhaps this effort will stimulate a more effective use of these silent tools in explaining faith-based concepts and constructs for new believers and those previously fully taught. Congregations are better educated than when my professional journey began seventy-five (75) years ago. Conceivably, more faith-based people will be openminded to a better understanding of crucial elements in composition. The preliminary feedback about Applied Mathematics and classic symbols suggests some may develop a use of these crucial tools to clarify meaning of faith-based concepts and constructs.

What is the importance of math? A simple answer: math makes life orderly and prevents chaos. Pure Mathematics searches for facts to produce data to solve basic problems that do not depend on human ingenuity but are governed by the longstanding rules of mathematics; Pure math may be for the work of academics. However, Applied Mathematics are resources for creative explanations related to the public square, marketplace and community and family life. Math may have unexpected applications to personal life and faith-based operations.

Applied Mathematics may influence more of human existence than most realize. As the world becomes more complicated, mathematics may become more informative for the future of the human race. Most academics do not agree on the difference between Pure and Applied math. Obviously, one aspect of math is speculative and theoretical and the exclusive domain of academics, while the other is useful when applied in real life situations. The study of math nurtures the power of reasoning, creativity, abstract or spatial thinking, more effective communication skills, critical thinking, and problem-solving abilities. These competencies are useful in most areas of human existence. **Why are Applied Mathematics not used to explain different aspects of sacred writing and faith-based living?**

What is the future use and understanding of Applied Mathematics and classic symbols in bringing clarity to interpreting faith-based expositions? The early readers of this book were Scholarly Referees; they read an early draft manuscript to review the concept.

Sidebar: What early readers said:

"A refreshing and awakening, a near thorough thesis of the path to moral excellence and eternal security. I urge you to pursue and complete. I also look forward to a teaching manual alongside this publication. You challenged me!

"I finished reading your masterpiece. You are an excellent writer with an impressive depth of knowledge in your wheelhouse. You are the master of metaphors, including those related to math and those already infused throughout the Bible. That may be a pastor thing. The content in each chapter reads like a sermon on steroids. While the message may not be new, it is your spin and style that give it meaning that's different. It forces the reader to pause and reflect. Your use of the math symbols for the metaphor or element for a chapter seems to work and isn't overpowering."

"The author, a gifted writer, uses math symbols and basic math principles to guide the reader to live a wholesome, God pleasing life and to experience God's love and forgiveness."

"This concept is mind blowing. My doctoral research makes a lot of difference now. Sure wish there was time to go back and perfect my statistics and use them as a normal tool in ministry."

"Great concepts and love the flow of logic it gives. Am loving the book.

Scholarly Referees

Paratan Balloo, MBA, DPhil, Vice Chancellor/CEO, OASIS University, Trinidad, West Indies
Ronald A. Berk, PhD, Professor Emeritus, Biostatistics & Measurement, Former Assistant Dean for Teaching, The Johns Hopkins University.
R. John Buuck, PhD, President Emeritus, Concordia University, (Wisconsin).
Alfred Cockfield, DPhil. Senior Pastor, God's Battalion of Prayer NETWORK, Brooklyn, NY.
Coy Webb, MBA, ThD, Family Pastor, New Life Church, Bluff City, TN.

After almost 90 years of struggling with the human journey, there remains hopeful anticipation that faith-based leaders will escape the prison of previous patterns and become better scholars using every item in the scholarly toolbox to assist those who want to understand, believe and behave sufficiently to walk the road less traveled and step by step maintain the spiritual journey toward moral excellence and find eternal peace in the Paradise of God.

Dedicated teachers with "lifelong learning" in mind taught me that *"All truth comes from God"* without respect as to where, how or from whom it was learned or experienced. They expressed the value of concepts, constructs, culture and language influence in the meaning of words and how a *prefix or suffix* assisted the understanding, word meaning and the emotion and/or movement they produce. An awareness developed how composition, integration, interpretation, translation, socialization, context, and the values of culture and tradition were woven into the fabric of life as a stabilizing force. All prior knowledge should be used in communicating the value of a message. Jesus taught fishermen how to use what they already knew to *"catch men alive"* as useful participants in the faith-based interchange.

Paul strongly urged new converts to *"abide in their calling (task, occupation, profession) and utilize their past experience in serving those they knew best before moving to the larger population."* (1 Corinthians 6:20-24) God's plans are to prepare one for service. Do not abandon your education, every jot and tittle can be of value in some aspect of life. All truth is of God and past experiences are designed to equip one for the future. In fact, all of life is about the future! The past is a memory, the present is only a moment, and the future is monumental with great importance and extends the lifespan and beyond.

Somehow an understanding of society as an aggregate of people living and working together in an orderly manner became a reality. This expanded my understanding to the structure of language, the construction and meaning of words. This provided a perspective on English grammar similar to the one developed in Seminary with *koine* Greek which assisted in translating the New Testament. This is why 42-years of my spare time were devoted to translating *koine* Greek into a rendering of common English normally read and spoken in the USA. Published as The *EVERGREEN Devotional New Testament (EDNT)*. The EDNT is available in 100 countries, on many websites, and on the Espresso Book Machine © worldwide or anywhere good books are sold (Hardcover; Trade cover; eBooks).

Through the years there has been a complication of the message of grace by the use of religious jargon and sectarian gobbledygook that has closed the ears of congregants and caused clergy to *"shoot at the moon and miss the woodpile."* Clyde Reid in his classic (1967) *The Empty Pulpit: A study in preaching as communication* declared *"... the American Pulpit empty because no one was listening."* Someone failed to inform the theologians that *"big words were constructed to think with, not to be use in communication.* This further complicates the faith-based message.

Based on education and experience during my first 75 years of life, following retiring from Oxford Graduate School (2007), my ministry through education continued through traveling, speaking and writing books. Based on my exegesis of the New Testament, education and personal experience, the attempt has been to pass on to others, things learned from good teachers, serious study and the pursuit of an integrated education. Born at the end of the Great Depression, suffering the death of my father at age 4, experiencing America's War years, growing up in the segregated South, pastoring churches, serving as a Military Chaplain during the Vietnam era, Chancellor of two graduate institutions, and serving as Professor of Education and Social Change for four decades, hopefully this my 58th book will plow some new ground in communicating treasured truth. With hopeful anticipation, Applied Mathematics, classic and historical symbols will be used to clarify aspects of the faith-based message in an attempt to answer the question, *Are All Believers Equal?*

$$[= := - \div x + \infty]$$

Historical Foundation

Peter wrote his second Letter (II Peter) to the same believers in Turkey about AD 67. The main themes of this letter were: an exhortation to spiritual growth; the necessity of holding on to truth; warnings against false teachers; and advice on lifestyle in view of the Lord's return.

A Common Privilege of Faith - 2 Peter 1:1-2

Simon Peter, a servant and an apostle of Jesus Christ, to those who share with us a common privilege of faith, justified as we are by our God and Savior Jesus Christ: 2. grace to you and peace, may it be multiplied to you through full knowledge of God and of Jesus our Lord.

Precious and Treasured Promises - 2 Peter 1:3-7

3. Since His divine power has bestowed upon us all things that are necessary for true life and true worship, through the full knowledge of Him who called us to His own glory and moral uprightness: 4. Since through these gifts He has bestowed upon us precious and treasured promises, you are to share the divine nature, leaving behind the corruption and passions of the world. 5. And you too have to contribute every effort on your own part, crowning your faith with moral excellence, and to moral excellence knowledge from books and teachers: 6. and to your knowledge self-control; and to self-control enduring steadfastness, and to enduring steadfastness godly worship; 7. and to godly worship brotherly kindness; and to brotherly kindness benevolent love.

An Abundant Entrance to the Kingdom - 2 Peter 1:8-11

8. Such gifts, when they are yours in full measure, will cause you to be neither unproductive nor unprofitable in the full knowledge of our Lord Jesus Christ. 9. He who lacks them is no better than a short-sighted man feeling his way about; and has forgotten that his old sins have been purged. 10. So, believers, be the more eager to confirm your calling and your choice: for if you do practice these virtues, you will make no false steps: 11. and you shall be richly supplied the entrance into the kingdom of our Lord and Savior Jesus Christ.

Refresh Your Memory - 2 Peter 1:12-16

12. It is for these reasons that I intend to constantly remind you of these things, although you know them well, and are grounded firmly in your memory; 14. the Lord Jesus Christ has showed me that shortly I must fold my tent. 15. Moreover, I will make it my endeavor that after my departure you will always remember these things. 16. For we have not pursued deceitfully devised folktales, but were eyewitnesses to His majesty when we made known to you the power and presence of our Lord Jesus Christ

Beware of Destructive Opinions - Peter 2:1-8

1. There were false prophets also among the people, and there will be false teachers among you, who secretly will bring destructive opinions among you, even denying the Master who bought them, and they will bring swift self- destruction. 2. Many will embrace their unashamed immorality and through them the True Way will be brought into disrepute. 3. And by greed with fabricated words they will make merchandise of you: their sentence was settled long ago, and now their damnation is not delayed. 4. Since God did not spare the sinning angels, consigning them to pits of gloom in a section of Hades reserved for punishment of the wicked until judgment. 5. And spared not the ancient world but guarded the eighth man Noah, a herald of righteousness, bringing a flood on a world of wicked men; 6. since He reduced the cities of Sodom and Gomorrah to ashes, when He sentenced them to destruction and gave an example of what happens to those who live ungodly; 7. and delivered righteous Lot, who was distressed by the immorality of lawless men, 8. for a righteous man to see and hear such lawless deeds was daily torment to his righteous soul.

Mathematical & Classic Symbols

In the ancient world, symbols were viewed with a sense of mystical fascination. Both religions and pagans used coded language and symbolism to express a primitive readable message. In recent times both public

Hopeful Anticipation...

and private symbols, logos, colors and shapes are used to assist recognition of messages requiring brevity or secrecy. This book uses common and mathematic symbols to support a concept or construct for better visual understanding and grounding.

[ ] ( )	Items to be considered first
<	Less than
:=	Equal by definition
−	Subtraction or negative
÷	Division of units
×	Multiplication or product
+	Plus or add together
∞	Infinity or unlimited number
>	Greater or more than
^→	Magnitude and direction
≠	Not equal
∵	Symbol for "because"
α	Proportional to something (resembles a fish)
±	Plus / minus used for range
~	Distributed (normal, negative or positive)
Σ	Sum or total
∴	Symbol for "therefore"
=	Equal or same value on both sides
⊕	Internal direct sum or memory

[.,?;':9 ( ) !" "…--_ /{ } @ [ ] *]

Punctuation marks and diacritical marks are symbols used to aid comprehension or the pronouncing of written language.

Sidebar: Punctuation marks in English grammar are symbols used to aid and clarify comprehension of written language. The New Testament was mostly Greek and without punctuation because the Greek language was specific enough for the reader to know what went together. When the translators rendered the Greek into English, they used the punctuation of English to further interpret the meaning. In reality, English grammar punctuation symbols are used to aid comprehension. Punctuation marks have different uses that enhance learning. The period, question mark, and exclamation point are **used to end sentences**. The comma, semicolon, colon, and dash **indicate a pause or break**. Parentheses **contain important words**, while hyphens **combine** them. Apostrophes show the **omission of letters**, and also show **possession**. There are 17 punctuation marks that are used in the English language. Punctuation marks have five categories: (1) Sentence endings: period, question mark, exclamation point; (2) Comma, colon, and semicolon; (3) Dash and hyphen; (4) Brackets, braces, and parentheses; (5) Apostrophe, quotation marks, and ellipsis.

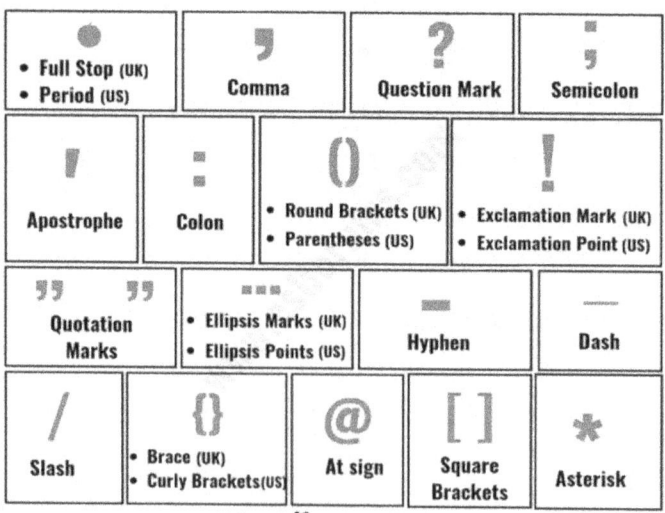

www.eslforums.com

NOTE: Since full understanding of letters and words alone are limited without punctuation and/or diacritical marks, why not add mathematics symbols to render or interpret language? However, punctuation may express intentional bias and/or diacritical marks may convey cultural expressions. Authors may show unintentionally their personal bias in the same manner. There is a difference in "word-for-word" translation and interpreting or rendering the meaning of a word, phrase, or unit of information to reach a current understanding. The use of symbols in this work is an attempt to enable better comprehension of the nuance of the composition and suggest subtle shades of meaning, feeling, or tone to improve comprehension.

ARE ALL BELIEVERS EQUAL?

THE POWER OF MORAL EXCELLENCE

(^→)

Introduction

Prior Learning Impacts Human Existence

(^→) Symbol indicates magnitude and direction

According to publication guidelines, an author should tell the reader something about the basic elements in the structure of a composition. This book presents "silent symbols" to enhance the meaning of a composition and demonstrate their significance by adding perspective to words, phrases and sentences **(e.g., math symbols, punctuation, diacritical marks, word origin, how a prefix or suffix influence meaning, etc.)**; all have value in writing. The use of these symbols could make difficult concepts easier to understand, more orderly and prevent chaos by increasing comprehension in all aspects of journalistic presentations. The study of math alone nurtures certain qualities, such as, the power of reasoning, creativity, abstract or spatial thinking, critical thinking, problem-solving competencies and better communication skills. Hopefully, the reader will see the value of both understanding and using symbols to improve their interpretation of faith-based concepts.

Geoffery Thomas, PhD, Head of Continuing Education and the Founding President of Kellogg College Oxford, UK, spoke for Degree Day at Oxford Graduate School in Tennessee. He was clear that America and England were separated by a common language. Also, we came to understand that the English write paragraphs and Americans write words. To further confuse me, Thomas explained that the English select books by authors and subject, and Americans read books by title with almost total disregard for authorship or publisher. This brought to mind a sad episode in my education: on a final university exam in a favorite subject a question worth 30 points was "Name the Author and Publisher of the Textbook?" Yes, I missed it, but learned a valued lesson.

During my academic career many trips were made to Oxford, UK to research at the Bodleian Library, the graduate library of the University. Many of my PhD students studying Education and Social Change or Leadership and Social Change have accompanied me to three world class libraries: the Library of Congress, the Bodleian Library at Oxford, UK and the British Library in London. This exposure to world class collections was structured to make students Lifelong Learners and increase awareness of cultural and traditional differences which adds perspective to one's worldview.

An English academic connected with Cambridge University, Constina Bicchieri, wrote a book, *"The Grammar of Society: The Nature and Dynamics of Social Norms."* Through a UK academic site, a PDF was secured for review prior to publication. Bicchieri examined social norms, fairness, cooperation, and reciprocity in an effort to understand their nature and dynamics, how they generated expectations, evolved and changed overtime. Drawing on multiple academic methods, including social psychology and experimental economics, she provided an integrated account of how social norms emerged, why and when they were followed, and where situations focused on relevancy.

Bicchieri's work stirred my academic juices as a Graduate Professor of Education and Social Change. The book thrust me into a deeper study of English grammar and linguistics, different aspects of sociology, social psychology, change theory and rekindled interest in my old math major. In particular, the origin of words and how various languages manipulate the formation of words and influenced the tracked changes of meaning. This increased my passion for etymology and historical-comparative linguistics.

> **Sidebar:** Language grammar includes syntax, morphology/ linguistics, and semantics. Syntax – includes the arrangement of words and phrases that form sentences and paragraphs, the grammar of a language which deals with prefix and suffix or an attribute as tense, mood, person, case, number, and gender. Morphology/linguistics studies how words are formed and has a perspective that the parts of language are living things and the relationships between structures and the changes of meaning. Semantics –a branch of linguistics and logic concerned with the meaning of words and phrases. One area of semantics observes the logical meaning, such as sense. reference, implication, and logical form. Another area studies the cognitive structure of meaning of words, phrases, sentences, or text as one entity.

James Grier Miller's classic *Living Systems Theory* (*LST*) (1978) and G. A. Swanson's applications of *LST* to accounting theory in his dissertation at Georgia Tech and his views on social research (1982) pushed me further into a fresh look at the integration of my academic studies, business and professional work experience,

Military Chaplaincy during the Vietnam era, my faith-based lifestyle, and ministry through education. Living Systems Theory pointed me to the integration and overlap of common ground in various subjects. Problem Solving based on Miller's classic (LST) has assisted in simplifying complex data across academic disciplines. Miller supplied a useful trichotomy of all systems; (1) **concrete** systems, (2) **abstracted** (selected) systems, and (3) **conceptual** systems.

Miller was concerned that various departments in academia did not interact with others and had their own private perspective on useful knowledge and the solution of problems. This was similar to leadership in the faith-based movement where each sectarian group believed they had the only access to the Holy Chalice which led to eternal redemption. Such exclusivity complicates education and weakens the whole concept of faith-based living and worship of One God Creator and Sustainer of the Universe.

Actually Miller's LST is a system to simplify thinking about complex problems (Swanson/Green, 1992) further explained the trichotomy. A **concrete system** is a nonrandom accumulation of matter, energy, and information in a region of physical space-time, which is organized into interacting, interrelated sub-systems or components. Examples of concrete systems are iron ore, a building, the human body, a computer or an organization. An **abstracted system** is a limited set of relationships *selected for sameness* by an observer. Examples of abstracted systems are a person's mental image of a lover, a modern theologian's assessment of Luther's theology, and the relationships such as length and angle existing on an object being measured and observed instrumentally. **Abstracted systems** may be sub-typed as transformable (*measured*) and non-transformable (*surrogated*) and are studied by **conceptual systems,** *a set of words, symbols, or numbers,* including those in computer simulations and programs, that have one or more, similarly ordered subset.

Although LST appears complicated, it is in reality a simplifying process to assist the overall understanding of a specific area of concern by multiple fields and professions. It facilitates interdisciplinary understanding and cooperation and builds on common ground and mutual exchange to better understand other disciplines and improves problem solving. Conceptual systems of *"words, symbols and numbers"* pointed toward Applied Mathematics and grammar paradigms in planning this book.

When analyzing a written manuscript there is a relationship between Applied Mathematics and grammar. Letters form words that symbolize *objects, action, or attributes.* In mathematics, numbers symbolize *amounts, patterns or relationships.* These words and numerical expressions create a conceptual system and a basis for improved focus on analysis of information and interpretation of data through critical thinking. As the world and human life becomes more complicated, it will be necessary to reach back into some original thinking and brazenly reach forward to new uses of old and workable ideas. The foundation of a paragraph is one idea fully developed. There must be a boldness in using creative ideas to go where others have refused to go because of tradition or the road less traveled.

Perhaps we should remember Roger's spaceship ventures or Dick Tracy's magic Wrist Radio. The cartoon originators were ahead of the great scientist of the time. Is it possible that modern education has forgotten foundational truths of early academics in their pursuit of solving problems of getting to the moon? Actually, NASA's space lab used LST symbols for critical systems to identify for quick reference each aspect of the space lab.

It is my firm belief that the silent symbols of mathematics and the classic symbols of history would expedite the awareness of individuals in a complicated world. Do you remember when public transportation vehicles had their route identified by numbers and colors for quick reference? My personal memory reaches back to a Chemistry textbook written after the first Atomic Bomb was used and the author attempted to feebly explain the force of the explosion and added *"and other God-like chemical reactions."*

The study of math nurtures certain qualities, such as, the power of reasoning, creativity, abstract or spatial thinking, critical thinking, problem-solving competencies and better communication skills. Taken in a strict sense, problem solving systems demand one to find (*identify, produce, construct, list, characterize*) all solutions. Taken in a less strict sense, the problem may not have just one solution. For practical problems dealing with something similar to the teaching and learning process or faith-based interpretation, the "strict sense" would have little value. The **context** and **cultural** identity of people must be considered.

In the ancient world, symbols were viewed with mystical fascination. Both religions and pagans used coded language and symbolism to express a primitive readable message. In recent times both public and private symbols, logos, colors and shapes are used to assist recognition of messages requiring brevity or secrecy. This book uses common and mathematical symbols to support a concept or construct for better comprehension.

The key to problem solving is found in the process of using centuries-old systems, structures and schemes related to Applied Mathematics and other ancient symbols used to relay a message. Such operations may influence more of human existence than most realize. As the world becomes more complicated and complex, the sophistication and simplification of mathematics, may become more informative for the future. Most disagree on the difference between Pure and Applied Mathematics. Obviously, one aspect of math is theoretical and the exclusive domain of mathematicians while the other is more practical and useful when applied to real-life situations. Mathematics has become a useful tool for critical thinkers. A current TV series (*Numbers*) uses the principles of mathematics to solve crimes. Why not apply some of these to faith-based constructs?

The interdisciplinary study of multiple subjects provides a useful tool for comprehensive learning and the integration of common subject matter in the overlap areas. In fact, studying only one subject greatly limits the ability to use common knowledge of one subject to better understand another. Differences divide and assists with perspective, while commonalities aid in connecting relevant facts to be placed in a conceptual system of words, symbols and numbers including those in computer simulations that have some similarity or sameness. This process provides insight into deeper knowledge of both subjects. A gradate professor came to me with a complaint, *"Some of my students are using what they learned in your class to answer my questions."* My response, *"Good, that is what graduate education is all about!"*

The more this happens, the easier one becomes sensitive to the math in music, the structure of grammar in language, the principles of logic, problem with punctuation in composition, the check and balance in problem solving, and the difficulty of translating from one language and culture to another and maintain the original intent of words. These elements provide insight to integration and interpretation of data from different academic fields. This also has implications for religion and faith-based considerations.

The legal profession, the field of medicine, and the theologians were restricted to education in one field of study. Then as these fields progressed, they specialized more and more until they whittled nothing down to a point and became stuck in the prison of previous patterns. Lawyers developed a specialized practice, medical doctors were either a generalists or a specialists, and those faith-based folk who concentrated in theology became sectarian and the practitioners were cocooned in a sectarian shelter and ventured out in their colorful vestments to attempt to breathe life into a decreasing congregation without reason or reality using recycled messages and platitudes with lost meaning and promises of a better afterlife without relevance to the present lifestyle.

While on the other hand, the medieval academic spent 8 to 9 years in studious pursuit of Arithmetic, Astronomy, Logic, Geometry, Grammar, Music, and Rhetoric. *[Take note of the overlapping structure and the connectivity through various levels of math.]* All this to provide tools to understand the past, present and future of human existence and the origin of the **cosmos**!

Law, medicine and theology were normally excluded from higher education because of a specialized body of knowledge from the past and a focused practice of required criteria for credentials in their social professions. At the end of this training they were placed in a special box and called on only when their specialty was needed. Doctoral-level studies were a scientific and exploratory research approach to higher learning for academics with theoretical and intellectual curriculum and objectives which differed greatly from the rote training of the social professions. The social professions received more and more of the same until reason and judgment replaced knowledge in many professions, and this was a limitation to most practitioners. They were restricted to dealing with personal problems of individuals and families while the academics tackled the process of advancing society but did not have a clear understanding of the problems. The interdisciplinary study of many subjects provides a useful tool for comprehensive learning and understanding the overlap of subject matter that provided insight to interpretation and integration of information from different academic fields. Most faith-based people and especially their leaders are stuck in the prison of previous pattern. Sorry, but no one is listening. All that remains is noise and clutter with no conviction or commitment to a moral lifestyle.

Consequently, the legal, medical and theological positions were limited by specialized education and focused training for competency in content areas of individual concerns, while higher learning curriculum prepared academics to solve the broader problems of a civil society. The social professions were provided limited skills for personal service and were aware of social problems but were not equipped with the social scientific tools for societal problem solving. The academics were provided scholarly, scientific and social proficiencies to deal with the broad public needs of a complex society but often did not sufficiently understand the problems. And so goes modern education!

This dilemma was solved by the integration of subject matter. All data contains error, but broader and more general study finds true facts and by using these facts from one subject in another subject a better understanding is realized. The principle of "going from the known to the unknown" was a great advance in the teaching/learning process. Have we forgotten Gregory's *Seven Laws of Teaching*? What about Bloom's *Revised Taxonomy* that includes six levels: remembering, understanding, applying, analyzing, evaluating, and creating. Perhaps it is time for some creative thinking about understanding and applying!

> **Sidebar:** Mathematics is the theoretical science of number, quantity, and space. But may be studied

as "pure mathematics" in its own right, or as it is applied to other disciplines as "applied mathematics." **What are the subjects and symbols in applied mathematics?** Here are ten procedures - (i) Numbers, Numerical Applications & Quantification (ii) Algebra, (iii) Mathematical Reasoning, (iv) Calculus, (v) Probability, (vi) Descriptive Statistics, (vii) Basics of Financial Mathematics, (viii) Coordinate Geometry, (ix) Faith-based Operations and Experience, (x) Social Problems and the process of Constructive social change.

Pure Mathematics searches for facts to produce knowledge to solve basic problems that do not depend on human ingenuity but are governed by the trusted rules of mathematics. Pure math may only be for the work of academics; and applied applications of mathematics as resources for creative explanations related to the public square, marketplace and group operations in the real world. However, all math may have unexpected applications.

Applied Mathematics attempts to model, predict and explain things in the real world by applying problem solving methods in different fields of human endeavor. Thus, Applied Mathematics is a combination of science and cognitive skills which could be used to clarify and simplify the concepts and constructs in sacred writings, benevolent institutions, compassionate organizations, and humanitarian enterprises. Adequately prepared individuals could connect the dots between the academic and the scholarly including systematic theology and practical faith-based behavior. This would enable moral leaders, through their capacity to solve real-world problems which may require mathematical and reasoning skills to develop new processes and procedures for social and faith-based difficulties in a civil society.

A problem that is not understood cannot be solved. To understand a problem, one must know the principal parts: the unknown, the facts or data, and the condition relating these facts to the unknown. Therefore, it is advisable in problem solving to pay close attention to the principal parts of the problem. The basic problem in understanding subject matter transfer is to link the detail facts of the information to the lives and culture of the learners. To transfer subject matter, one must find a way to excite and direct the self-activity of others and tell them nothing they can experience for themselves. Learning is a cooperative venture. Guidance from books and teachers, together with personal commitment to rigorous study, form a conceptual system that supports the advanced learning process.

An existing difficulty must be analyzed and classified. Take time to study the facts involved. Every problem has three components: (1) an unknown, (2) something known or given (data), and (3) a condition which specifies how the unknown is linked to the data. The condition is an essential part of the problem. What kind of problem is this? If a problem can be classified and recognized as a type, progress has been made toward finding a solution. The method learned to solve this problem must now be recalled. A good classification system should suggest the type of problem and the type of solution.

There are two general types of problems:
1. To complete what is missing: **to find**
2. To support a hunch or confirm an assumption: **to confirm**

The aim of a problem "to find" is to *(construct, produce, obtain, or identify)* whether a certain assertion is true or false, to affirm or reject it. When one asks, "How can I find a way?" a problem to find is presented. Yet, when one asks, "Did he find the way?" a problem to confirm is presented.

Problem to Find: The goal of a problem to find is to locate a certain object, the unknown of the problem, satisfying the condition that relates the unknown to the data. A problem clearly stated must specify the category to which the unknown belongs and the condition the unknown is to satisfy. Is the unknown a teaching method, a word to reach the heart, or an action to deal with a moral or ethical problem?

Problem to Confirm: The principal parts of a problem to confirm are called the hypothesis and the conclusion from the tentative assumptions. To support the proposition, one must discover a binding link between the principal parts, the hypothesis and the conclusion. The hypothesis when tested by an appropriate statistical process should support the conclusion.

What are the facts in the data? What does the problem solver understand? Can the nature of the problem be expressed in common language? Can evidence be given for the facts? May the facts be applied to real life situations? These are **problems to confirm** and must be supported by evidence produced by the solver. Whoever takes these questions seriously has presented a **"problem to confirm."** The facts of the process should be reproduced in a real-life situation.

Sidebar: The Task Force that initiated Oxford Graduate School as the American Center for

Religion Society Studies (ACRSS) offered a praxis Doctor of Philosophy which integrated sociology and religion. The emphasis on social scientific research had two basic difficulties: (1) members of the social professions normally feared any form of mathematics. And (2) mature adults needed a **both/and** residency that divided the semester into class time and field research using their profession as an applications laboratory.

The problem-solving aspects of applied mathematics and the tested field of statistics were essential to the social research goals of the program. It was decided to offer Statistics as a language for academic transfer of subject matter. Surprisingly, it worked! Doctors, lawyers, clergy, counsellors, teachers saw the value of scholarly communication through statistics. After all, statistics is the sophisticated use of high school and early college math; once the rules are learned it appears less complicated and the resistance is weakened. Viewing statistics as a language, a means to communicate, enhances a "value added" issue to graduate study and encourages the use of Referred Journals and peer reviewed research to report tested hypotheses and use statistically supported findings. Consequently, Statistics satisfied one of the two languages required for the Doctor of Philosophy and opened the door to the second area of concern.

With a workable approach to social scientific research, the next academic hurtle was a modified residency that met the needs for advanced study of mature adult learners. A **both/and** method. The academic background of mature learners with experience in a professional field opened the door for a modified residence. Semester-length Terms were divided into Campus Residency and Field Research. All class-based assignments and academic writing projects were completed in a virtual professional field-based applications laboratory. It was a **both/and** approach that satisfied both the UK-European and the US criteria for Graduate study for mature well-educated adults.

Social scientific research with statistics as a language and a program of developmental readings in Referred Journals with peer reviewed reports together with learning logs assisted meeting the other basic difficulty with a praxis approach to class-based credit and a modified residence graduate currency in their profession and the continued use of peer interface. This prepared lifelong learners and enabled post- review research as a path to constructive social change. In fact, those who earned a Doctor of Philosophy were told: **"GO CHANGE THE WORLD!"**

President Lincoln's Emancipation Proclamation (1863) dealt only with the inequality of enslavement in the Confederacy. And in 1865 the effort in Congress to Amend the Constitution needed a way to break the gridlock. Lincoln used a rule of *mathematical reasoning* in Euclid's 2000-year-old book on *mechanical law* to facilitate passage. Euclid's self-evident notions included *"Things which are equal to the same things are equal to each other."* This was clearly **"equal by definition"** (:=). Another influential self-evident notion was *"What is true of the whole is not true of the part."* This suggested "not equal" ($\neq$) and the logical and mathematical converse gave elected officials cover to vote for the 13th Amendment and for all States except Delaware and Kentucky to vote for Ratification.

Dr. King's assessment against injustice and discrimination was based on a mathematical construct "equal by definition" (:=). He reasoned that the rule of law was not in line with valued governmental documents: "We hold these truths to be self-evident: that all men are created equal, that they are endowed by their Creator with certain unalienable rights, that among these are life, liberty, and the pursuit of happiness." In Jefferson's draft, the words were "sacred and un-deniable," but Franklin edited the draft to read "self-evident" and it was approved by vote. This is how elected statesmen solve differences – through reasoning and the voting process not public rebuke or denunciation. All constructive change must be assessed, evaluated, lawful, non-violent and the work of mature elected officials.

(<)

Grounding:

Grafting into the Spiritual Rootstock

(<) Less than

12. It is for these reasons that I intend to constantly remind you of these things, although you know them well, and are grounded firmly in your memory; 14. the Lord Jesus Christ has showed me that shortly I must fold my tent. 15. Moreover, I will make it my endeavor that after my departure you will always remember these things. 16. For we have not pursued deceitfully devised folktale but were eyewitnesses to His majesty when we made known to you the power and presence of our Lord Jesus Christ. (2 Peter 1:12-16 EDNT)

The function and value of a graft is less than (<) the host rootstock. The character and productivity of the grafted branch is limited by the nature of the process. The rootstock is grounded with a taproot and a supply system of roots that provides lifegiving nourishment to all and has self-sufficiency to live and be productive without the graft, but the grafted branch is dependent on the host for both life and productivity. The faith-based rootstock has a taproot growing vertically downward for grounding and supplies nourishment for the whole tree.

On the other hand, a grafted branch is a parasite obtaining most of its nutrition from the host rootstock without contributing to the benefit of the host. A graft as a parasite depends on the host for grounding and nourishment for both foliage and fruit. In this regard a grafted branch is less than (<) the host. The graft is dependent on the roots of the host for both life and fruit. The growth of a grafted branch is upward toward the sunlight and the necessary rain from above while the roots of the host supplies grounding and life to the graft.

In faith-based operations individuals and groups become attached to the historic rootstock which has a taproot going back to Abraham's One God Faith. While the host is providing beneficial nourishment, the graft takes from the host without giving back. The graft uses sources for life received from the host to grow its own foliage and fruit and normally expresses itself by a showy difference.

The graft is selfish and is **considered a "taker" rather than a "giver"** into which individual and groups are grafted reaches for grounding in ancient history. The roots of Monotheistic faith reach back to Abraham and the concept of One God and Father of all Creation. The New Testament "pristine way" that eventually became known as Christianity.

> **Sidebar:** In my Grandfather Green's yard was a small red apple tree, but one of the branches produced an apple yellowish in color. As a curious boy, grandfather explained that the branch producing the yellow apple was grafted into the rootstock of the red apple tree. My big question, "How do you do that? How does it work"? With great patience grandfather explained the process of splitting the branches in a specific way and placing them together. Then binding them tightly with twine and covering the whole joint with bee's wax. He shared that the grafting process is a "taker" and not a "giver."

Judaism worshiped the One God of Abraham, the early Messiah-like people or Christians saw the One God relationally with three distinct observable relationship qualities that formed the undivided holistic character of God, expressed as Father, Son and Spirit. Later Islam, also a monotheistic faith, saw Allah as the One and Only God, Creator and Sustainer of the Universe. Regardless of the term used to identify the Supreme Deity; Jehovah, Jesus Christ or Allah or any other acknowledged surrogate expression; it is self-evident that all who serve One God should be working to provide for all His Creation. All three monotheistic religions have an ethic of reciprocity as to the proper way to treat others. This is the Golden Rule. Yet, some converts do not measure up to the faith of Abraham.

> **Sidebar: The Monotheistic Ethic of Reciprocity**
>
> **Judaism** – "What is hateful to you, do not do to your fellowman. This is the entire Law; all the rest is commentary." (Talmud, Shabbat 3id)
>
> **Christianity** –"As you would that men should do unto you, do you also to them likewise." (Luke 6:31)
>
> **Islam** –The ethics of reciprocity is an Islamic moral principle which calls upon people to treat others the way they would like to be treated. Although it is not mentioned in the Quran, the principle was stated many times by the Prophet Muhammed.

Although the **taproot** of pristine Christianity goes back to New Testament times, it also reaches back to Abraham, with **branch roots** of smaller size and little grounding to which others pray and give allegiance. These roots have shallow depth with little nourishment and are of about the same size. Then there are the grafted transplants that grow somewhere other than the

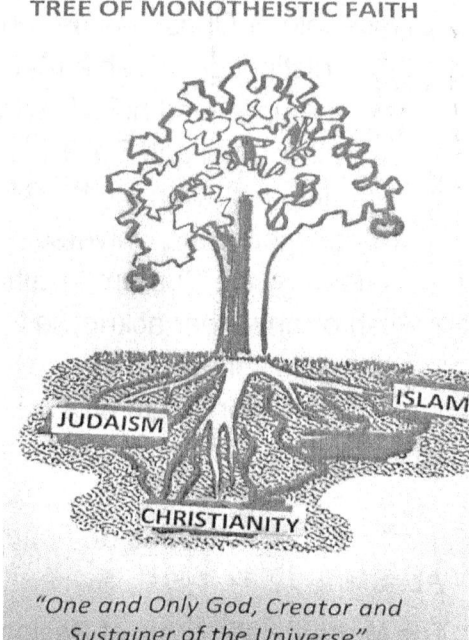

TREE OF MONOTHEISTIC FAITH

"One and Only God, Creator and Sustainer of the Universe"

root system with only indirect grounding and become interlopers and suckers that extract the life out of the host's rootstock. Such grafts are less than (<) the pristine rootstock.

> **Sidebar:** Rootstock in grafting deals with the unseen roots that sustain the plant. Actual grafting takes place in a plant with a well-developed root system. Another difficulty is when a vine cutting is used for rootstock, after a short time it may form a single entity. After a longer period, it is difficult to detect the graft, but the product always contains the properties of genetically different plants. This is where the complexity of grafting into spiritual rootstock shows the differences in various branches. All sectarian groups who claim connection with a historic faith-based heritage are not the same and produce different fruit. "By their fruit you will know them."
> In fact, all living organisms have negative or slow

growth early in their lifespan. There may be equality by definition (:=), but the differences are obvious with little commonalities. Just as newly born infants all do not grow and develop at the same rate or develop to the same degree of maturity. All living organisms need tender watch care during the early stages of their lifespan.

Since Abraham believed God, Monotheism has been deeply rooted in ancient human history. These facts must be understood. *The process of moving from lost to found or from conviction of sin to conversion by faith is not the same for everyone.* Some make a drastic faith leap forward and their life is drastically changed, and all things become new as their step-by-step journey begins on the straight and narrow way that leads to life eternal. Others give mental assent to truth without a contrite heart and are hesitant to walk the pathway that leads to a moral lifestyle. They stumble and sidestep off the path of grace and truth and become identified by brand name only. They are Catholic, Protestant or Baptist, Methodist, Lutheran, or some other sectarian brand. This is where the math symbol (:=) *"equal by definition"* enters the legal and theological conversation. Euclid's general truism, *"What is true of the whole is not true of the part"* and the converse ought to be considered in the calculating equality: *"what is true of the part is not true of the whole."*

> 11. And He gave some to be messengers, and some preachers, and some missionaries, and some teaching pastors; 12. for the ultimate purpose of equipping the saints for the work of serving, for the building up of the body of Christ: 13. until we all attain the same faith, and the experiential knowledge of the Son of God, unto mature manhood, unto the full measure of development

in Christ: 14. that we no longer behave as young children, driven before the wind of each new teaching, by the trickery and sneakiness of men, whereby they ambush with deceitful schemes; 15. but arriving at truth in love, you may grow up into Him in all things, Who is the head, even Christ: 16. from whom the whole body is in harmony and compressed together by that which every joint supplies, according to the increase of every part, so that the body is increased within itself in love. (Ephesians 4:11-16 EDNT)

A grafted branch takes three to eight weeks to heal and does not produce fruit for one to three years. This timetable cannot be rushed and suggests that new people introduced to a different culture or civil society will need special attention in the early weeks and continued watch care for one to three years to become acclimated and assimilated into a new way of life. This is true of a new relationship, marriage, workmate, neighbor or faith-based change. To avoid unintended errors, the mature among us must take the initiative in assisting the adjustment of new folk in our sphere of influence. Who is your new neighbor? What can you do to facilitate their adjustment and adaptation to a new environment without causing damage to their self-image or self-worth? Are you willing to try? If not, you are part of the problem.

Nature nourishes trees with a supply of a watery liquid called sap. It moves upward in part through an intricate supply system. It seems that the grafted branch sucks nourishment from the rootstock of the host but does not give anything back. Other branches assist the tree through sap-lifting forces created by evaporation and transpiration. This is a kind of breathing water through the leaves. A grafted branch does not participate in this process. The grafted branch uses all the sunshine and rain on its leaves to produce its own special fruit. This is the nature of all living organisms: their viability as an entity is primary.

> *16. When the first loaf is holy: the whole batch is consecrated: and if the root of the tree is holy, so are the branches. 17. And if some branches were broken off, and **you being of a wild olive tree were grafted into the tree and became also a partaker of the root and richness of the olive tree;** 18. you must not look down on the branches that were broken off, but **remember you do not support the root, the root supports you.** 19. You may say, the branches were broken off that I might be grafted into the tree. 20. But it was from lack of faith that they were broken off, and you stand in their place by faith. **Stop being proud but be on guard:** 21. for if God spared not the natural branches, pay attention lest He also not spare you. 22. Consider both the kindness and strict justice of God: on them who fell, strict justice; but toward you, kindness, **if you continue steadfast in His goodness; otherwise, you will also be pruned from the tree.** 23. And should they not remain in unbelief, God is able to graft them in again. (Romans 11:16-23 EDNT)*

When the concept of grafting is understood, new folk are given a time for processing adjustment. It is a strain on the original unit, because the grafted branch actually takes away from the other aspects of fruit bearing. Too many grafts could suck the life out of a tree in the process of sustaining an alien branch and different fruit. An adjustment here requires common sense and fairness. A new culture may soon overpower the dominant culture, and everyone loses. The secret is *com/promise* (this is not a bad word it simply means *"together with/ promise."* To achieve agreement, sameness must be found with all parties. All significant change requires a contract-type agreement where each side cooperates and gives up something to gain something. There is no constructive change without understanding the construct of **com/ promise!**

Scripture in Romans 11 explained a self-centered perspective, which resulted in "taking without giving." To illustrate the concept, Paul used the character of a grafted branch and said, **"You do not support the root, but the root supports you."** As Jim Elliot, one of the five missionaries killed in Ecuador (1956) wrote **"He is no fool who gives that which he cannot keep, to gain that which he cannot lose!"** This is the mindset that readily opens the door to constructive change. A grafted branch does not support the root, which nourishes its life, but the root supports the grafted branch. The graft may live, grow, produce foliage and even fruit, but remains an unorthodox part of the larger unit: *This is the tie that binds people together in the trap called "codependency." In faith-based entities, when too many take from the rootstock without giving back a morally changed lifestyle the main unit is weakened. A certain, but unknown, number of takers will cause a slow decline and the prune* the "takers" is almost nonexistent in most faith-based groups.*

____*The set-back of pruning and the necessity of the process is discussed in Chapter 34 of WHY CHURCHES DIE (1972, 2007) ISBN 978-0-9796019-0-3.

Humans seem to manipulate others for their own self-interest, even when they appear to be engaged in useful behavior. Just as a grafted branch may be aggressively selfish and become a liability to the original unit, a disruptive child, a new neighbor, or a recent faith-based convert may overburden their new connection. A graft may also become a hindrance to growth and fruit bearing by sapping strength from the rootstock. In this regard authorities often fail to grasp the unintended consequences of attempting to integrate multiple cultures or attempt to assimilate too many differences that weaken the dominant culture and at times damaged a society beyond restoration. This is why all such efforts of deliberate interchange should be thought out by mature folk and a workable plan developed for the benefit of all concerned based on social scientific research without the mudslinging of political opinions.

A lesson was learned about givers and takers: and most of us at some time or the other may be on both sides of that equation. Takers should work at giving and givers should not begrudge their giving. A pastor of a small church in a West Virginia coal camp, who had 12 children of his own, was asked, "Brother how do you make it with such a big family and a small church?" Pastor Little's answer was clear *"Every little helps!"* Scripture is clear *"It is more blessed to give than to receive."* Why? The receiver may still have needs while the giver out of overage gives with some left. The coinage value of a gift is always measured in the cost to the giver. What does the giver have left? Remember the widow and her two small coins? Jesus said, *"She has given more that all others because she gave out of her poverty?"* This is when little is much when God assigns the coinage value of a gift.

> (See The Children's Bread (2018) – Unlocking whole life stewardship and Accessing faith-based economics and personal wealth, ISBN 978-1-935434-90-0)

The Tree of Monotheism includes these examples and grafts into these systems are not equal neither is the fruit or product the same. A rose by any other name may have a good fragrance; however, a man-made faith-based system not connected to the historic rootstock of One God is a counterfeit simulated copy. Paul writing to the Corinthians about a supreme commitment to the truth of the Gospel shared *"less by any means, when I have preached to others, I myself should be rejected as a worthless coin."* (I Corinthians 9:25-27 EDNT) Since Paul considered that such could happen even to him, what about the false teachers who with intent leapfrog over the truth to deceive others?

> *6. Let no man mislead you with words devoid of truth: because these things bring the anger of God upon the disobedient. 7. Do not associate with such things. 8. For once your heart was in darkness, but now it is filled with light from the Lord: behave as the product of light: 9. (for the. product of light is seen in all goodness, righteousness, and sincerity;) 10. be living proof of what is well-pleasing to the Lord. 11. And have no friendship with the activities of darkness, but rather admonish them. 12. For their secret actions are too disgraceful to even talk about. 13. But all things that are censured are made obvious by the light. (Ephesians 5:6-13 EDNT)*

The whole process would be much easier if everyone understood the problems of grafting. An individual or a group cannot long take from the nourishing roots of a common culture without giving back their unreserved loyalty and support for the cause of unity in both institutions and the public discourse. Parents, families and faith-based folk must be caring and compassionate humanitarians advancing individual and family welfare and constructive change for all in their sphere of influence. Some of my early books were about this issue: *

> ____* Why Churches Die, (1972, 2007) Why Christianity Fails in America, (2010) Remedial and Surrogate Parenting –Human development 0-20, (2013) and Tear Down These Walls (2013). Newer books: The Power of Forgiveness and Reconciliation, (2020) Beyond Pulpit, Classroom and Lecture Hall (2021), and Navigating Multiculturalism (2021) also speaks to the premise of this book.

The Commission (Matthew. 28:16-20) was not a command to "go and do," but a challenge to "do as you go!" Believers were told to wait for the power of the Spirit at the HARVEST FESTIVAL. The Spirit was to provide guidance to effectively follow-up those who received the teaching, embraced the teacher, and became an active learner. Why would scholars turn a participle (as you personally go) into an imperative (command to "go")? All academics and scholars are influenced by an intellectual and cultural bias that takes years and firm effort to overcome.

PART TWO: First Steps

STEP I: START WITH COMMON FAITH (+)
STEP II: BUILD ON PRECIOUS PROMISES (∵)
STEP III: SHARE BENEVOLENT LOVE [= :=−÷× + ∞]
STEP IV: MULTIPLY BROTHERLY KINDNESS (×)

ARE ALL BELIEVERS EQUAL?

THE POWER OF MORAL EXCELLENCE

(+)
Step I
Start with a Common Faith

(+) Plus or add together

Simon Peter, a servant and an apostle of Jesus Christ, to those who share with us a common privilege of faith, justified as we are by our God and Savior Jesus Christ: 2. grace to you and peace, may it be multiplied to you through full knowledge of God and of Jesus our Lord.

(2 Peter 1:1-2 EDNT)

Peter was writing to justified believers or beginners (+) that God's grace and peace should be multiplied **(x)** through personal knowledge gained from spiritual experience in their relationship with God and His People. From day one New Testament converts shared a common faith but knew it was only the first "small step" on the journey to an afterlife with God. Throughout scripture the New Covenant was clear that the old system where God winked at the lack of knowledge and bad behavior was cancelled. Now since the Sacrifice of Jesus, all who were drawn to God by the Spirit were given a measure of faith and convicted of sin and with a contrite heart accepted forgiveness. This covered the past and believers were encouraged by the multiplying of grace and peace to start the walk of faith. Just to start a new journey was not sufficient to balance the spiritual equation. **Starting** also required **departing** from the previous path of sin which leads to perdition and walking on the narrow path which leads to an afterlife in God's Presence. Starting is the first step and all who endure the hardships of the faith-based journey will be assisted by a loving God.

Some attend a local church with no intention of starting the journey on the narrow pathway of faith. They only seek immediate influence or personal gain. Attending a religious service might assist their business or they would find new friends. The purpose may be to influence a potential mate that they are honest and moral in their dealings with others. Perhaps they were invited by a friend and were curious. Some become a churchgoer to please their family not to seek redemption or prepare for the afterlife. A false start in a race usually means a selfish motive to get a personal advantage over others. Among the *"easy believe-ism"* section of church-ology, some attach themselves to a faith-based congregation to soothe their conscience that they are trying to do better without a contrite heart or an intention of changing their lifestyle.

A false start creates false security and does not predict a completion of the journey. No one comes to God without the drawing of the Spirit. The invitation of a friend may create an awareness of the need for a life-change or the person my go through the process of joining a church and attending a *"class for new converts"* and even submit to public baptism without a genuine spiritual experience that leads to regeneration and change of lifestyle. This may increase the numbers in the congregation, but it is doubtful it will change the population of Heaven.

How are faith-based attributes multiplied **(X)** in the lives of true converts? Spiritual characteristics are increased in believers in two ways. The Greeks had two words for knowledge: (1) knowledge gained from books and teachers, and (2) full knowledge or the experiential realization based on personal awareness. This is "first-hand" learning from direct contact with the compassionate love of God in answered prayer and observing God working in one's own life and the lives of others. Learning the reason for a situation being different than expected or desired is an example. The reality of seeing the wisdom of parents assists in seeing the guiding Hand of God pointing in the right direction or enabling one to better understand what God is doing in their life. Jesus said to His close followers *"You have not chosen Me, but I have chosen you."* There is a spiritual place on the journey when believers cease to be just servants of God and develop an *"intimate friendship with Jesus"* and this brings better understanding of God's working in their lives. This is full or experiential knowledge gained by a personal relationship or transactional knowledge of divine action.

> *15. I no longer call you (servants) or bond slaves because a bond-slave does not know what his Lord does: but you I have called friends; for all things that I have heard of my Father I have made known to you. 16. You have not chosen me, but I have chosen you, and appointed you to go out and bring in fruit, and that your fruit should remain and that you should obtain answers to your prayers to make **(x)** them fruitful.* (John 15:15-17 EDNT)

In a mathematical equation there is equality or balance on both sides. The problem-solving dichotomy exists here. A problem to "find" must be "confirmed." In conversion there is the confession of sin and the profession of faith as a confirmation. Faith must be accompanied with works to achieve fulfilment. A gift must be received. A wrong must be forgiven. Bad behavior must be *subtracted* from a believer's lifestyle. Good behavior must be *added* to a faith-based life. Spiritual fruit must be preserved. A new convert must be nurtured and guided into to the straight and narrow way. Those who endure to the end of the journey will have an abundant entrance into God's Kingdom. The plan that produced salvation included great sacrifice by Jesus; to balance the equation there must be clear evidence of perseverance on the journey to reach the ultimate place God has prepared for those who truly love Him and complete the journey. **The end is worth the journey!**

In the present age of entitlement, it is common for many to seek an easy way. The young and the able bodied who will not work but demand special privileges. It boils down to *pay without work, free rent without responsibilities, keys to a car with money for gas, free association without accountability;* in other words, they want the milk but will not buy the cow, and demand a free pass to life, comp tickets to entertainment, and everything NOW that it took their parents 25 years to acquire.

The adults who complain about this situation are part of the problem. There was no *"tough love"* no *"straight talk"* no *"get a job and buy it yourself!"* All this includes the abdication of parental responsibility for their own children. What happened to *"earn your own money"* or *"honest work builds character"* or each *"tub must set on its own bottom."* The Garden edict *"Earn your bread through hard work"* has not been rescinded. **Are we all caught in the trap of codependency?**

A careful analysis of sacred writings would show that all converts at the beginning of the spiritual journey have an equal opportunity to utilize the gift of faith and the knowledge from books and teachers to grow in grace and experiential knowledge to gain full spiritual maturity. The difference appears to be how the measure of faith divinely bestowed on all seekers is daily used to walk the lighted pathway. It is obvious that some *"stay too close to the getting-in place"* and fail to take the logical steps needed to move forward on the journey. Sadly, they settle into a false security because of "easy believe-ism." As Paul would say, "They become a *worthless coin*," (of no value to the Congregation). God assesses the coinage value of consistent steps on the faith journey. Conversion is not home plate or a rocking chair by grandma's fireplace, one must cover the bases to score a victorious run for the home team. One may be the weak player on the team and in a must win situation the team must bypass the weak one and "switch batters." The opportunity to shoulder the bat for the winning run is an earned privilege. All may be on the team and are (:=) equal by definition but not equal in performance.

Most still believe that individuals are given some "measure of faith" and the common privilege to use this faith to increase the quality of their life. Sacred writings by Paul listed three things that would endure forever: **faith, hope, and love.** He designated this by using a forward-looking form of *agape love*, as the greatest of the three. This is **benevolent love**, *doing what the one who loves deems needed by the one loved,* and reflects God's willful direction toward all mankind. Faith and love are given to each and every member of the human race and becomes the starting point for the faith-based journey with the guidance of mature family and friends.

Provided an individual takes the first step of faith with a contrite heart and accepts forgiveness she/he is adopted into the Family of God as a newly born infant and must grow and develop over time into faith-based maturity. When conversion is compared with the human birth cycle one begins to understand why converts were classified as being "born again." Relating a "new birth" with conception, fetus attachment to the womb, development and the live birth of an infant, the series of steps required to achieve faith-based conversion and spiritual maturity are better understood. A brief study of conception, the early germinal process, embryonic development and the fetal timetable for viability could bring clarity for those who work with converts and the young and growing disciples. (See Remedial and Surrogate Parenting (2013). ISBN 978-1-9354344-81).

Faith is multiplied by this benevolent love and God's blessings to encourage growth in faith-based living. God's gift of enduring faith is shared by all and brings an long-lasting hope when adequately applied to real-life situations. To have hope one must have both desire and expectancy. With a hopeful desire and the expectancy of faith, a convert may push aside fear and move forward with benevolent love to grow in grace and divine blessings. These are the blessings of the beginning, but the end is worth the journey (joined by family and friends.)

Conversion is not a static state or stationary position. Hitting the ball and standing on home plate does not score for the home team. The word itself has a Latin suffix **(ion)** that clearly shows a required process and/or action. It appears conversion assumes that individuals are equal, (:=) *"by definition"* not by behavior or maturity. The quality of conversion may depend on the individual's action and reaction with the process of regeneration and transformation that comes with genuine conversion. There is a process of personally sustained action steps following an initial act of faith: this is not conversion, but the process of conviction. In the early days of *juris prudence,* a convict *"agreed with the verdict and asked for mercy."* This is what happens in true conversion: one convinced of a sinful life based on agreement with God's Word develops a penitent heart and seeks redemption. When faith and action are sufficient, the work of conversion begins. It is a process not a static situation. Leaders do new seekers a disservice when they practice "easy believe-ism" and fail to follow up with scriptural guidance on how to live a godly life in this present evil world.

First comes heart-felt conviction, then repentance, then the acceptance of forgiveness, then the process of regeneration of the inner spirit, afterwards comes reformation through the steps toward moral excellence in lifestyle. Steps must be taken forward in a new direction. From sacred writings ***con/version*** appears to mean *"**together with**"* a changed *"**version**"* of personal behavior that manifests in a change into a moral and ethical lifestyle. A "version" of the same person emerges *"wearing the full armor of God"* **standing on divine**

promises not sitting on the premises. Now a convert is ready to start the faith-based journey on the pathway to an eternal future.

This change is observable by others. All this suggests an individual approach that makes each one accountable and responsible to make the necessary adjustments to lifestyle behavior that is apparent to others. New converts must not only appear as a believer, but also perform and emerge into a distinctively changed individual. The changed life of a convert becomes a testimony of God's love and grace to family and friends (enemies, too).

The big question: are all believers equal? Equality is a deep and complicated subject on which few actually agree. Any stipulation of equality itself is not exactly equivalent or identical to all others. There are different social, moral, legal, and personal perspectives. These are based on *"equal by definition, equal under the law, equal by creation, equal by existence, equal by habit and lifestyle, equal as observed by others or equal by self-evaluation."* These are selfish attitudes and the only items acceptable are those which bring benefit to the person regardless of how it may impact others.

Many years ago, Euclid of Alexandria, a mathematician, did some critical thinking about equality. He developed some general truisms and presented them as self-evident wisdom:

3. Things which equal the same thing equal one another.
4. If equals are added to equals, the sum is equal.
5. If equals are subtracted from equals, the remainders are equal.

6. Things which coincide in nature, character or function equal one another.
7. The whole is greater than the part.

Euclid's final concept, **"the whole is greater than the part"** together with the converse seems most informative on the subject at hand. Some in every cohort or group membership will not measure up to expectations. Although the parts make up the whole and what is true of the whole may be an achievable objective of the parts; however, in a faith-based operation, when an individual does not measure up to group standards, often a biased tolerance permits those who lag behind to maintain a connection. Obviously, what is true of the whole is not true of each part. There is no responsibility without accountability. The group may be strong, but similar to a chain: it is really no stronger than the weakest link. The same would be true in a cohort of faith-based converts. The "weak ones" must be nourished quickly before they become discouraged and their behavior limits the influence of the whole group. A strong chain may have a weak link, or a wagon wheel may have a less than perfect spoke. The same would be true in a group of faith-based converts. This weakness must not be overlooked and must be confronted with the individual.

Euclid's general truisms ought to be considered in the calculation of equality. A careful comparison of sacred writings would show that all who start the journey have an equal opportunity to utilize the knowledge from books and teachers and use the measure of faith divinely bestowed on all seekers. This would also reveal that some "stay too close to the getting in place" and fail to take the logical steps needed to move forward on the journey. Some in every cohort or collective group membership will not measure up to expectations. Although the parts make up the whole, what is true of the whole is not always true of each part. [All disagreement on this point will be settled by God on Judgment Day!]

From my experience, military commands have two parts: **preparatory** and **execution.** It is clear that the command *"to the rear --March!"* includes a preparatory element, an order to halt, to change direction with a new destination. The process and/or action included an authoritative order to quickly turn about and go in the opposite direction at the same pace of 90 steps a minute. If an individual or a platoon hears the command and some do not follow the clear order, anyone could see the breakdown of discipline. They are all soldiers by definition, but some were not following orders and continued to travel in the wrong direction. This clearly caused disorder in the ranks.

There is no standing at the edge of the Water of Life with a fear of getting wet; converts must take the plunge and move forward by faith to demonstrate full obedience to divine order and discipline. The process or change initiated by conversion is completed by personal action-steps in a different life direction than was previously contemplated. It is a new way of life with different traveling companions and a group of new friends. Normally one does not learn to swim without getting in the water.

Being identified with a group does not make one equal to the quality standard of the entity. There may be similarities but there will always be differences. A failure to follow through, at times called negative participation, is usually the roadblock to progress. In faith-based issues this is obvious. The new and the young must have a time of becoming, i.e. growing in grace and knowledge) to be recognized as a team member instead of a bench warmer. This is where the real dichotomy is observed: some are participators and others are spectators. Without focus and singleness of mind and heart, it is difficult for most to walk the straight and narrow way or climb the Hill of Difficulty that appears just beyond the next side-step on the journey. Although the end is worth the journey, many are distracted by their human struggles and stumble off the path and need a quick step to keep from falling.

8. Now God is continually able to overflow you with self-sufficiency always making you competent to pour out to the good of others: 9. as it is written, His generosity is scattered to the poor; His love-deeds are never forgotten. (2 Corinthians 9:6-9 EDNT)

Some may be in the fight, part of the struggle, but faint-hearted and discouraged and are unequal in courage, strength and steadfastness. They are blessed with forgiveness of the past but fail to persevere and endure the hardships of the journey. All should remember the struggle of early believers walking on the way struggled to navigating through a maze of Judaism and Greco-Roman mythology and paganism. Yes, they were attempting to walk on the way established by Jesus and His Disciples, but there were trials, tribulations and persecutions. These early saints survived by holding on to the promise that God was Present with them "always" even to the end ....of the journey.

19. As you personally go, (going) therefore, and make disciples of all nations, baptizing them in the name of the Father, and of the Son, and of the Holy Spirit: 20. teaching them to observe all things whatever I have commanded you: and behold, **I am with you always, even unto the end of the world.** *Amen.* (Matthew 28:19-20 EDNT)

Sidebar: Traveling alone down Interstate 75 south of Atlanta, the long journey ahead suggested a hitch hiker might be good company. As the young man entered the car, "Good morning, my name is Hollis Green; I am a Christian." The response was "Carl Krudof, I am a philosopher." The young philosopher was baited, "Do you write your philosophy down, or do you just talk?" He claimed to write important thoughts down. Carl was asked about his most recent writings. He said, "I have just written a definition of God, but I don't believe there is one." [A definition of God by a philosopher who doesn't believe in God. This was going to be interesting.]

Reaching into the back seat to retrieve a small unzipped notebook, Carl began to read: "God is the singular, possessive, abstraction of the adverb." [He is a philosopher; my teachers talked that way. He was asked to repeat the first statement.] He repeated, "God is the singular, possessive, abstraction of the adverb." [What's an adverb? I've been out of school too long.] Carl continued, "An adverb is the linguistic manifestation of a life process."

It is my conviction that God provides both the situation and the supplies to share what others are seeking. The discussion centered on Carl's definition. It was good theology for a philosopher who did not believe. This was discussed at length. His use of the present tense initiated a long exchange. The singularity of one God was discussed. The possessive nature of God was considered. God's ways being past finding out seemed to be an abstraction.

Somewhere in Carl's intellectual comprehension, the use and function of the adverb was the key to an adequate perception of God. "An adverb is the linguistic manifestation of a life process." God was not viable to Carl because there was no systematic order relating the words and symbols about a Divine Person to his personal reality. God in this case was the big Verb, and Carl had never witnessed the action of God in real time. He needed someone who had personally experienced the power and action of God to adjust the semantics and syntax of the experience to a language he could accept. Carl needed the same touch of experiential reality that Thomas of scripture desired. He needed a touch of first-hand personal reality. He required a manifestation of the resurrected life of Jesus Christ. At last, the course of action was clear. Carl needed to see one of God's adverbs.

This called for a new introduction, **"Good morning, my name is Hollis Green; I am one of God's adverbs."** A spark of cognition ignited; Carl's mind was open; his heart and mind were ready, and the Holy Spirit had done His work. A simple walk down the Roman Road of scripture brought Carl face to face with the reality of the man Christ Jesus. He accepted not only the present tense existence of the Creator, but a personal relationship with Jesus, the Son. Carl was greeted as a Brother, a fellow adverb to go forth and magnify the essence of God as a redeemed adverb.

What the world needs is more of God's Adverbs sharing the excitement of living the faith-based life. Are you ready to be one of God's Adverbs and modify and describe clearly the action of God in your life?

(**See Appendix B**—"God is…")

Sidebar: This side of Heaven, it appears impossible for the human mind to fully comprehend God's infinite and awe-inspiring nature. In the Bible, however, The One God has shared enough truths about His Nature and distinguishing characteristics referred to as "Father, Son, and Holy Spirit." In (Genesis 1:26) God speaks, "Let us make man in our image, in our likeness…" God views man as a wholistic entity with all parts intimately interconnected and explainable only by reference to the whole, because the whole is greater than the part. In geometry, an equilateral triangle has three equal sides. So it is with The Creator God who exists in three coeternal and consubstantial persons as Father, Son, and Holy Spirit! **(See Appendix B and C** – How a personal God communicates)

HOW A PERSONAL GOD COMMUNICATES

GOD communicated with man from Creation to the Birth of Jesus as the FATHER.

FATHER

GOD communicated with man from the Birth of Jesus to the Ascension as the SON.

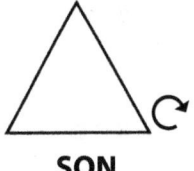
SON

GOD communicated with man from the Harvest Festival (Pentecost) to the present as the HOLY SPIRIT.

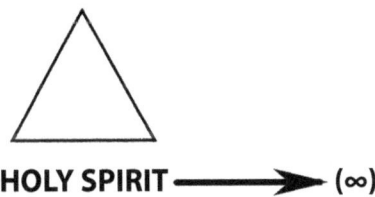

ARE ALL BELIEVERS EQUAL?

THE POWER OF MORAL EXCELLENCE

(∴)

STEP II

Build on Precious Promises

(∴) *Symbol for because*

3. Since His divine power has bestowed upon us all things that are necessary for true life and true worship, through the full knowledge of Him who called us to His own glory and moral uprightness:
4. **Since through these gifts He has bestowed upon us precious and treasured promises.** *You are to share the divine nature, leaving behind the corruption and passions of the world. 5.* **And you too have to contribute every effort on your own part,** *(2 Peter 1: 2-5 EDNT)*

Because the foundation stones for the walk of faith are placed as secure stepping-stones for the journey, and identified as *faith, peace, love, grace, joy, and hope,* be it known to all that not far into the journey hardships, sufferings and troubles will be encountered. Why? Because God has confidence in those who start the journey by faith. These difficulties produce *patient endurance, character, more hope and trust in God's promises.* The Greeks called this process the path to full or experiential knowledge that comes with working through personal difficulties and trusting a Higher Power for solutions. Because of such testing new believers gain strength, courage and more benevolent love provided by the Holy Spirit as they witness God's intervention in their life and the lives of family and friends. Conversion separates one from sins and provides the courage to witness to friends. Sharing the new life is part of the power to persevere on the journey of faith.

> 17. You really don't think I was irresponsible, do you? Do you think that my plans were according to human impulse that with me there was yes, yes, and no, no? 18. But as surely as God is true, my word had no equivocation or mixture of yes and no. 19. For the Son of God, Jesus Christ, who was preached among you by us, even by me, Silvanus and Timothy, was not yes and no, we never wavered between yes and no. 20. For to all the promises of God we supplied the affirmative yes, and in God is the final Amen, through us (so be it) to the glory of God. 21. And now He who stablished us with you in Christ and has anointed us with the Spirit is God; 22. Who has also sealed us and given a down payment of the Spirit in our hearts. (2 Corinthians1: 17-22 EDNT)

This basic trust in God is strengthened by the experiential knowledge that God has given to each contrite soul a degree of faith to be used incrementally as they grow in grace and knowledge and travel step by step on the narrow way that leads to life eternal. This limited quantity of faith is bolstered by a major dose of God's one-way love placed in the heart and soul of those who believe. The Divine Love is one single dose which produces in the heart and life of the faithful certain virtues, identified as the Fruit (singular) of the Spirit which generates the byproducts of *agape* love: **Increased faith, gentleness, goodness, joy, long-suffering, peace, self-control and tolerance.** (Galatians 5:22)

The foundation stone is **benevolent love,** the next step leads to **brotherly kindness.** Then **godly worship** brings **enduring steadfastness** and **self-control.** A giant step forward brings **knowledge from books and teachers** which identify the final steps to **moral excellence** which is the **Crowning of a Believer's Faith** that began with **benevolent love.** True faith produces **generosity and compassionate caring for other believers and the disadvantaged.** These are the steps the Holy Spirit guides each believer to moral excellence and an abundant entrance into the Afterlife. The process is never over until it is over at death's door.

> *It took just a week to make the moon and stars,*
> *the sun and the earth and Jupiter*
> *and Mars. How loving and patient*
> *He must be Cause He's still*
> *workin' on me.*
>
> **Joel Hemphill**

Yet, many swimming in the human cesspool of misfortune and misery have little hope that things will change for the better.

In the bosom of all mankind, beats a heart full of hope for an improved human existence and an afterlife in a better place. Most are stuck in the prison of previous patterns and cannot find a way of escape. In the midst of despair, life is hit-and-miss, pitch and pass, and searching without finding. It is simply trial and error daily until all hope dries up and no one cares enough to provide encouragement or even a smile.

Then when there seems to be no way forward or upward, all hope sinks into the desolation of isolation. Such brokenness opens the door for a divine encounter that would drastically change things provided the person was willing to listen for truth, reach for faith and claim God's blessings. When an individual realizes that all good things come from God, they are ready to listen. *"God is not willing for anyone to perish but that all should come to the knowledge of the truth."*

> *If you can't get someone off your mind,*
> *Pray for them. You may be the only*
> *One that cares enough to do so.*

When the hand of faith reaches down to lift a discouraged soul and enable them to see beyond their trouble, God is beginning His redemptive work. Simple faith in God can clear the vision sufficiently to see those things that hope brings. Yet, a group of seekers are made up of such a diverse field that no two are exactly the same. They have some sameness, but their differences are obvious. God has made of one blood all nation and humans have many of the same characteristics and difficulties. As long as there is life there is hope for a better future. When faith rises, hope begins to work. Faith is the evidence of things desired and provides evidence of things not yet seen. That is the basic "faith" which God gives to every human to use for their betterment. This diverse group is difficult to define. In social research it becomes an artificial dichotomy and can be scientifically studied on the basis of sameness. In other words a cohort of converts from the unconverted pool may be only (:=) equal by definition, meaning they have some sameness, but there are significant differences.

The cohort becomes an artificial dichotomy when past behavior or present circumstance places them in a cohort box identified as Group One and Group Two. It is artificial because the observer does not determine who is in each group; individuals by their behavior place themselves in a particular group. **Group One** gives mental assent to faith-based material and say a few correct words and participate in baptism, membership, and some sacraments, but the *"word was not woven into the fabric of their faith."* Because of this false start they become discouraged and fail to walk the path of peace and hope.

> *The good news was proclaimed to us, as well as to them: but the word was not heard and therefore did not profit them, because it was not woven into the fabric of their faith when it was spoken.* (Hebrews 4:2 EDNT)

Group Two consist of those who have a contrite spirit and are truly sorry for their sinfulness, by faith they accept forgiveness and acknowledge that the process of conversion that initiates regeneration is working in their life. They acquire a desire to grow in grace and knowledge and begin to develop a new lifestyle and to share fellowship with believers and seek a place to worship where the things of God have value. They begin to take consistent and constructive steps on the straight and narrow way. They focus their broad-mindedness to the small stepping-stones which begins the upward journey. Once they climb the first Hill of Difficulty, they become fully aware that their faith in God has nullified their past, made them rejoice in a redeemed fellowship and continue to gain knowledge from books and teachers that assist their journey and brings more faith and hope for the future. Now they understand that eternal life in God's presence is in their future. They are convinced the end is worth the journey and desire their family, friends, and neighbors to share in God's true blessings of grace.

> *12. So, then, lift up the drooping hands, and the weak knees; 13. and plant your feet in a straight path, lest someone who is weak stumble out of the path; but be restored to strength instead. (Hebrews 12:2-13 EDNT)*

Both groups desire eternal life but some fail to accept God's forgiveness with a contrite heart and refuse to permit the Holy Spirit to work the process of change together with a new version of their life. Conversion and regeneration are the work of the Holy Spirit, but the convert must be willing and cooperative with the process and accept changes in their behavior and lifestyle. Conversion is a clarion call to develop conduct and character consistent with faith-based living. Peter and Paul both saw a future departure from true faith in the last days.

> *11. For the saving mercy of God has appeared to all mankind, 12. Instructing us to deny all wickedness and worldly desires and to now live discreet, honest, and God-fearing lives in the world. 13. Looking for the blood purchased hope, and the magnificent appearing of the great God and our Savior Jesus Christ; 14. Who gave Himself for us, that He might ransom us from all wickedness, and purify a people as his personal treasure*, eager for good deeds. 15. Speak encouraging words about these things and admonish with authority. (Titus 2:11-15 EDNT)* _____
>
> * v14 The Greek word used for possession was "periousious" from the present participle feminine of a compound [peri] and [eimi] meaning "being beyond usual, i.e. special (one's own). Yet the KJB translators chose to use the word "peculiar" from the

> Latin [peculium] meaning one's own property or in Roman law meaning "private property". Although it did not essentially change the meaning for English readers because "peculiar" was commonly used in 1611, but this translation became a problem later as some individuals used the concept of appearing peculiar as a standard for the inner cleansing of holiness.

The human race shares a precious gift of faith given as an endowed capacity to both believe and behave God's Word as presented in scared scripture and proclaimed by called messengers to those who will listen and learn. This process is under the influence of the Holy Spirit and produces benevolent love. Because of this love believers are declared righteous by faith and enjoy the peace of God. This divine presence remains with believers as they take steps of faith on the straight and narrow way. New travelers on the way soon learn that it is difficult to be too broad minded and progress on the narrow path toward moral excellence in mindset and lifestyle.

Through the righteousness of God there is no partiality for believers. All who begin the journey have God's promises that they will be sufficiently equipped for all the rough patches along the way. My mother frequently shared *"When God sends you on a difficult path, He always arranges for better shoes."* Then she would quote, (Ephesians 6:15) *"Having your feet shod with the preparation of the Gospel."* The foreshadow of this assurance were the six (6) Cities of Refuge in the Old Testament. Three (3) were on each side of the Jordan River to provide quick and easy access. The roadway to these cities were required to be straight with a smoothed surface, so no one would stumble seeking refuge.

Bridges were built over ditches. All accused falsely of crimes could find refuge in one of these cities until they could receive an unbiased hearing. Even in the Old Testament, God made provision for those seeking refuge from false accusers. (Numbers 35:1-15; Deuteronomy 4:41; Joshua 20:1-9) Now under the New Covenant, true seekers are received with open arms and all with a contrite heart are forgiven and shown the way to walk on the stepping-stones of promise *"with their feet fitted with the knowledge of the gospel of peace"* that enables them to negotiate the difficult parts of the journey that leads to Eternal Life. Although the Holy Spirit becomes a guide for the journey, believers need the knowledge of sacred scripture to negotiate the difficult parts of the pathway.

Personal diligence is able to make your calling certain and your choice sure. Faith and knowledge are common doors to spiritual promises. Grace and peace are multiplied as one willingly walks through open doors and steps firmly on God's promises. He has provided in the endowment of grace everything for a moral lifestyle and godly worship. Not only must a seeker be willing to receive, they must be diligent in the pursuit of the steps that lead to moral excellence as a witness of God's transforming power. There is no partiality with God's gifts, they are available to the whole human race. This is not a "wait-and-see entitlement;" God's promised gifts are for those who diligently seek to be established in the Family of God and are willing to actively share the witness of grace with diligence. This is a lesson that must be learned! *It will be on the final exam!*

> **Sidebar:** My Grandfather Green gave me a job during hay cutting time. The barn was on a hill, and I was supposed to run along behind the wagon,

loaded with hay, and scotch the wheels when the mules needed a rest. I placed several large stones up the hill to the barn. After several loads of hay and several scotching episodes, I became tired. Seeing the coupling pole sticking out behind the wagon, I decided to straddle it and ride awhile. Forgetting my job, the mules got tired, stopped, and the wagon began to roll back. Finally, with Grandfather's commands the mules were able to stop the backward movement of the wagon with a jerk. With the jolt, the hay slid off the back of the wagon and pushed me off the coupling pole.

Grandfather dug me out from under the hay, and reminded me, **"You can't just ride the coupling pole, son, you must scotch the wagon. That was your job!"** The understood lesson was clear: one can't just take a free ride without understanding the accountability for assigned tasks. When one fails to do assigned tasks, it makes more work for everyone. And sometimes it creates circumstances that produce serious consequences.

I never forgot my coupling pole ride, my "fall from grace," or my **lack of diligence in following instructions.** The feeling at the bottom of a load of hay on a rocky hill road, with the sound of a wagon, two mules, and Grandfather above. When you know it is your fault, the lesson is a hard one to learn. Surely, there would be extenuating circumstances: **"I was only a little boy, the hot day and hard work was tiring, I had done well on several trips up that hill, I needed a rest..."** However, when you know it was your fault, no excuse would work. Such lessons stick with you all your days, nights, and weekends. The worst part of my lack of diligence was that I disappointed my grandfather who always expressed such confidence in my ability to accomplish things. This failure of discipline was a learned and remembered lesson and required a serious work ethic to compensate for my lack of diligence.

Are All Believers Equal?

The Power Of Moral Excellence

$$[= := - \div \times + \infty]$$

Step III

Share Sincere Benevolent Love

(=) Equal in quality to God's unqualified love for everyone.
(:=) Converts are only equal by definition, not in lifestyle.
(−) Converts must subtract negatives from their lifestyle.
(+) Believers contribute every effort to add value to their faith.
(÷) Believers divide their time and talent with others.
(×) Believers multiply their love by sharing God's love.
(∞) Believers follow God's steps to an abundant afterlife.

Since the love of God has been poured into our hearts by the Holy Spirit (Romans 5:5), believers must share God's love equally **(=)** without prejudice or purpose of evasion, sharing God's Love as opportunity affords.

At first a seeker is only a convert by definition (:=). Next, they must begin to subtract **(−)** things that are not compatible with a faith-based lifestyle. Then they must add **(+)** to their faith, suitable behavior that will demonstrate their profession and add validity by an observable lifestyle. The initial act of forgiveness deals with the past. Next, a convert must concern themselves with the present evil world and the sinful nature of humanity and permit regeneration to work by submitting to the guidance of the Holy Spirit.

A profession of faith does not change a human being into a Heavenly Angel singing around the Throne of God beyond the sinful and polluted world. The human nature remains intact and must be controlled by submission to the Holy Spirit. Submission is not giving up being a member of the human race; it is simply organizing life under different rules. Just the same, salvation does not separate one from their friends only their sins. As a member of the global human race, they still live in the same community and the same house, have the same family, the same friends and workmates. They still live among the unrepentant and unforgiven ones who remain captured and controlled by the forces of evil. Following initial conversion; however, the same person has a choice to use the faith and benevolent love to voluntarily submit to the leadership of the Spirit and organizes their life around those things which please God. Prior to conversion an individual was under the control of the baser compulsive drives of a sinful nature with little choice of behavior. Such a one is pushed into an unrestrained lifestyle that leads to progressive debauchery. A true believer has the power to choose and is told in James 1:2 *"rejoice because you have a choice."* Surely God does not tempt one to sin but does test a believer's commitment to choose correctly.

> *12. Blood-related* and fortunate is the man who flinches not under the enticement of testing: for when he is proved trustworthy, he shall be given the wreath of honor that verifies vitality, which God promised to all who worship out of a benevolent heart. 13. Let no man say when he is enticed, God allured me to evil: for God does not use wickedness to validate the trustworthiness of any man. 14. But every man is attracted to wicked deeds, when he chooses action based on personal desire, and hope of pleasure. 15. Then when personal desire has joined together with enticement, it produces a voluntary transgression: and this offense produces separation from observant morality, and at the end*

separation from God. 16. Do not wander from the right or deviate from the true course, my cherished band of brothers. 17. Every unspoiled and true benefaction is from above, and comes down from the Father of all light, with whom there is no changeableness, neither a dark side where there is no light. (James 1:12-17 EDNT)

> * v12 to bless, in Middle English used at the time of KJV, had a meaning related to "blood" used to consecrate an altar; thus, the use of "blood-related."

The symbol for division (÷) suggests believers should divide their time, talent, blessings, and burdens. Sincere conversion causes not only a change of attitude but of action, one does not change their attitude by simply changing their mind: there must be positive action to change attitude, which is a predisposition to behave! Time must be divided between work and worship, family and friends, and the present and future. Talent includes ability, resources, tithe and gifts and all must be put to constructive use. A division of wages would include tithe, love gifts and resources divided between family care, operational expenses, family legacy, care for the poor and humanitarian endeavors. Blessings benefit others. Burdens must be divided; a burden shared is not as heavy and believers are encouraged to *"bear one another's burdens"* and to *"cast their cares on Jesus."* A failure to divide time, talent, blessing or burden will create an atmosphere conducive to stumbling which in scripture is *"an offence or an enticement to sin,"* and in human terms a stumble may be overcome by a quick step forward to avoid a fall. One may easily see that these issues are not equally shared by all who profess to be on the step-by-step journey to moral excellence.

The multiplication symbol (×) in the new life of a convert relates to simultaneously increasing the attributes of a follower of Jesus toward others. In sharing with staff in a childcare facility, one worker asked, *"How do I divide my time and love between ten children assigned to my care?"* The answer was simple *"Multiply your love and your time."* Her response *"How do I multiply my love and time?"* Again, the common-sense answer is best: *"If children have the same need, problem or behavioral issue, group them and deal with several at once. This will multiply time and personal care! And love is multiplied by showing no partiality but treating all the same."* Remember, God's benevolent love is continually shared by the Spirit to all believers walking in fellowship with Jesus. As a believer one does not lose the quantity of love but increases it by multiplication.

All who believe and are changed in heart and lifestyle will enjoy an abundant entrance into the Kingdom of God. The symbol (∞) for *infinity or unlimited number* represents those who have an abundant entrance and positive reinforcement for the future and the afterlife promised to believers. The future journey will include *"The present troubles of this life are not worthy to be compared with the future blessings."* (Romans 8:18)

God's benevolent love is freely available to all who would accept His free gift of faith as a foundation upon which to build a faith-based lifestyle. After the initial gift of God's love is accepted by an individual and faith takes root in the heart, a situational audit happens in the human soul as one realizes the Word is True and they are outside looking into the sanctuary of redemption. A sense of lostness settles into a troubled soul as the Holy Spirit begins the process of shedding God's love into the essence of an immortal soul. That spark of living essence breathed into Adam in the Garden and passed via blood through the human race becomes aware that a dying human is in the presence of a Living God.

Then the inner most being cries out to a loving God for forgiveness of the past and a fresh start on the new and living way. How does a believer get more of God? This is a misnomer! One cannot be more spiritual than at the moment of a true conversion when all past sins are forgiven, and one stands justified before God. It is not getting more of God, but God getting more of the individual. The whole mistaken idea about the Holy Spirit sharing God's love is that some actually believe the person is getting more of God when in reality God's purpose is to possess more of the individual. In reality, it is through submission to the Holy Spirit that God has more control of the individual though sharing His benevolent love and equipping the convert for the journey ahead. In turn believers *"pay it forward"* and share God's love with others and this increases the value of their witness and extends the message of grace.

As believers line up under God's strategic plan for their life, the individual is influenced more by the Holy Spirit. True redemption comes only from God; however, growth in grace requires the convert to submit to walk the straight and narrow pathway that leads to life eternal. It seems that the human equation by nature more easily yields to the flesh than to the Holy Spirit. This makes the equation unequal and grieves the Spirit because when human nature controls an individual, they easily fall prey to temptation and stumble on the narrow pathway. A stumble means a quick step is needed to keep from falling. This requires effort on the believer's part. And assistance by other believers is always helpful.

> **1. Brethren, if a man should make an unintended error due to weakness, you who are regenerated, repair and adjust him with a teachable spirit; continue considering yourself, lest you also be tempted to make a false step.** 2. Practice in sharing the heavy burdens of others, and you will fulfill the principle of Christ. 3. If a man supposes himself to be something when he is really nothing, he deceives himself. 4. <u>Let every man test himself for innocence, and then he shall rejoice in himself and not in another.</u> 5. **For every man must carry his own personal load.** (Galatians 6:1-10 EDNT)

> **16. Acknowledge your failures and side steps one to another, and pray for yourselves and for one another, that you may be made spiritually whole again.** When a righteous man prays fervently there is great power in prayer. 17. Elijah was a man similar to us and he prayed earnestly that it should not rain, and for three years and six months no rain fell upon the earth. 18. And he prayed again, and the heaven gave rain and the earth put forth her fruit. **19. My band of believers, if any of you do stray from the true path, and one turn you about, 20. let the brother know, that he who turns one back from the error of his way into the right path, covers many faults and makes him safe, restoring his usefulness to the congregation.** (James 5:15-20 EDNT)

This is why forgiveness of the past is not enough, there must be the process of regeneration which reignites the flame of life breathed into Adam when man became a living soul. The Holy Spirit is present to assist this process, but the seeker must **submit**; that is, *line up under new arrangements with guidance from the Spirit.* The Old Testament points to man's conscience as the "lamp of the Lord searching the deepest part of the human heart and prompts change." (Proverbs 20:27)

> *30. Never distress the Holy Spirit of God, whereby you have been marked for the day of redemption. 31. Let your bitter frame of mind, anger and violent outbreak or brawling, and abusive language, be put away from you with all hatred: 32. Become gracious to one another, tenderly affectionate, ready to forgive one another, even as God for Christ's sake forgave you.* (Ephesians 4:30-32 EDNT)

When the Holy Spirit shines scriptural light on a believer's pathway and exposes wrong behavior, it is time to ask forgiveness. Yes, conversion took care of past sins, but wrong behavior requires a penitent heart to maintain a current relationship with God. After the initial submission to the Presence of the Spirit, there is a sense that the Spirit is always present to spread the compassionate love of Jesus on all weak ankles and wounded knees. The Spirit convicts of sin, righteousness and the coming judgement. This acts as a prompt for confession of sin and enables one to receive more caring compassionate love into their heart and life. It emboldens the public witness to the profession of faith. The Spirit guides those who willingly line up under divine authority and follows the lighted pathway. Without the Spirit's light the narrow path is dark and stumbling and more darkness soon follows. A stumble followed by a quick step of faith brings new strength for the journey. Human weakness and doubts cause carelessness, and one fails to be watchful of obstacles in the path. Human nature works against all spiritual progress because the *ego* wants to be in charge. The conscious mind is part of the human identity and is considered the *"self."* This conscious self gets in the way of spiritual progress. However, in Mark 14, Jesus returned from praying in the Garden and found His disciples sleeping because they gave in to human weakness and went to sleep while Jesus prayed!

> *37. And He returned, and found them sleeping, and said to Peter, Simon, did you fall asleep? Could you not keep watch one hour? 38. Watch and pray, in case you enter into temptation. The spirit certainly is willing, but the human body is without strength.* Paul a mature believer still had problems with his human nature: *25. And every man who enters the race practices rigid self-control. They do this to win a wreath that will soon wither, but we seek a crown that will not fade. 26. I run but not aimlessly; so I fight, but not as a shadow boxer: 27. but I beat my body black and blue and bring it into subjection: lest by any means, when I have preached to others, I myself should be rejected as a worthless coin.*
> (I Corinthians 9:25-27 EDNT)

Walking on the streets of Oxford, England with my son, Barton, we saw graffiti scrawled on a college wall *"Life is not a paragraph."* As a father/teacher, Bart was asked what it meant; his answer was clear, *"A paragraph is one idea fully developed, and if you are alive then you are not fully developed... so life can't be a paragraph."* Human development is incremental progress over time, and one must have a guiding purpose to stay on the determined course despite distractions and difficulties.

> *14. The seed is the word. 15. And these are the ones by the hard path, where the word was sown; but when they heard, Satan immediately came and took away the word that was placed in their hearts. 16. And these are the ones sown on stony ground who, when they heard the word, immediately receive it with gladness; 17. but they had no real roots, and so endure for only a short time: afterward, when affliction or persecution came for the word's sake, they were easily hurt. 18. And these are they who are sown among prickly weeds such as hear the word, 19.and the cares of this world and the deceiving pleasures of riches, and the craving for material things enters in and chokes the word, and it becomes unproductive. 20. And these are they who received the word on good ground such as hear the word, and receives it, and becomes productive, some thirtyfold, some sixty, and some one hundred-fold. 21. And Jesus continued, Is a lamp put under a bushel, or under a bed? And not placed on a lampstand?* (MARK 4:19-21 EDNT)

The foundation stone is **benevolent love.** True faith produces **generosity and compassionate caring for other believers and the disadvantaged.** These are the sequential steps the Holy Spirit guides each believer on the journey to moral excellence *...crowning your faith with* **benevolent love;** *and* **brotherly kindness;** *and* **godly worship;** *and* **enduring steadfastness;** *and* **self-control;** *and* **knowledge from books and teachers:** and **moral excellence.**

> *1. Since we stand declared righteous by faith, let us enjoy the possession of peace with God through*

our Lord Jesus Christ: 2. by Whom we have access by faith into the grace wherein we stand, and we rejoice in the hope of God's glory. 3. And not only, but we also glory in hardships and sufferings: because we know that troubles produce patient endurance; 4. and **patient endurance** *approves* **character;** *and character brings* **hope:** *5. and a faithful trust in God's promises will never put us to shame; because* **the love which God has for us is poured out in abundance in our hearts by the Holy Spirit which is God's gift.** *6. For when we were yet powerless, at God's time, Christ died for the ungodly as a finished deed.* (Romans 5:1.6 EDNT)

25. Wherefore speak every man truth with his neighbor without falsehood: for we are members bound one to another. 26. Have righteous anger without sin: let not the sun go down on your anger: 27. Neither give an opportunity to the devil. 28. The thief must steal no more, but let him do honest work with his hands, that he may have something to share with the needy. 29. Let no unwholesome words come from your mouth, but only good words for enriching, that it may serve as a blessing to the hearers. Jam25. Wherefore speak every man truth with his neighbor without falsehood: for we are members bound one to another. 26. Have righteous anger without sin: let not the sun go down on your anger: 27. Neither give an opportunity to the devil. 28. The thief must steal no more, but let him do honest work with his hands, that he may have something to share with the needy. 29. Let no unwholesome words come from your mouth, but only good words for enriching, that it may serve as a blessing to the hearer. (Ephesians 4:25-29 EDNT)

(×)
Step IV

Multiply Brotherly Kindness

(×) Multiplication

21. Since we have a high priest over the house of God; 22. let us come forward with a sincere heart crammed full of faith, having our guilty consciences purified by sprinkling, and our bodies washed with pure water: 23. let us not waver in acknowledging the faith, we profess, we have a promise from one who is true to His word! ***24. Let us keep one another in mind, always ready with love and acts of piety, 25. let us not abandon our meeting together, as some habitually do, but let us encourage one another, and all the more as we see the great day drawing near.*** *(Hebrews 10:21-25 EDNT)*

To ensure balance in life, believers multiply (×) their God-given benevolent love and divide (÷) their time, resources and talent between family and friends, private devotions and public worship and work and recreation; however, multiplying spiritual love to others is based on opportunity and explicit instructions in the Word. Part of the diligence to add to faith by sharing God's love is related directly to "brotherly kindness." This is an attitude *"a predisposition to behave in a certain manner."* All should remember that such as time, talent, resources and opportunities are limited and must be impartial by multiplying love until all have a sufficient share. On the other hand, benevolent love is resupplied by the Holy

Spirit and is not limited by sharing. In fact, benevolent love is multiplied as it is used to spread faith, hope and divine love that forgives the past, adds love to life in the present, and brings hope for the future of others. When the meal in the barrel of faith is mixed with the oil of the Spirit and shared with others it is not diminished but multiplied to meet present and future needs. (1 Kings 17:16)

> *19. My band of believers, if any of you do stray from the true path, and one turn you about, 20. let the brother know, that he who turns one back from the error of his way into the right path, covers many faults and makes him safe, restoring his usefulness to the congregation. (James 5:19-20 EDNT)*

> *8. Finally, you must think the same thoughts, share difficulties with one another, having automatic interdependence with brotherly kindness; be tender-hearted and humble-minded: 9. you must not repay injury with injury, or hard words with hard words, but bless those who curse you. For you were called to give kind words to others and come to a well-spoken eulogy at the end. 10. For the one wishing to love life and see prosperous days, let him avoid an evil tongue and cunning words. 11. Habitually avoid evil and do good; let him seek and follow peace. (1 Peter 3:8-11 EDNT)*

The divine gift of **benevolent love** naturally prepares one for sharing affectionate and cordial kindness to others. Not only sharing loving kindness with family and friends, but also to neighbors, strangers and enemies. This **brotherly kindness** is to be multiplied within the Family of Faith as well as shared with the world outside the walls of worship. Such attentiveness and persistence will draw believers to fellowship and **godly worship** which demonstrates the "worth and value" of God and worship with the Family of Believers. True and able-bodied believers will be consistent in fellowship with togetherness in godly worship. This connectedness through worship of God with other believers will bring **enduring steadfastness** and **self-control** which greatly enhances a consistent witness to the world. A faith-based believer must become a diligent student and an excited learner to gain practical **knowledge from books and teachers:** and the experiential knowledge which comes from participating in persistent prayer and active listening to God's daily working to meet the needs of His People. When this process is completed, individual believers and the band of worshipers will be crowning their faith with observable **moral excellence.**

The process began with conviction of sin and a contrite heart which led to forgiveness and receiving by faith God's benevolent love. God generously bestowed a measure of faith and showed the path to full fellowship with God. Believers started with God's forgiveness and benevolent love which was shared through brotherly kindness. Then godly worship and enduring steadfastness and self-control were added. To grow in grace and knowledge the process continued through knowledge gained from books and teachers and finally crowned by moral excellence demonstrated in observable lifestyle. All who faithfully continue the journey on the straight and narrow path will ultimately receive an abundant entry into God's eternal kingdom. Only then is the "finished work on Calvary" individually completed in the lives of believers. Then Mark's "double 13" is fulfilled: *"And you shall be hated by all men for My Name's sake:*

but he who endures to the end, the same will be saved." (Mark 13:13). This supports the pristine dogma known as **"Perseverance of the Saints."**

The foundation stones were **faith and benevolent love** which produced **brotherly kindness.** Then **godly worship** brought **enduring steadfastness** and **self-control.** Growth in **grace** and **knowledge came from books and teachers** and identified the steps to moral excellence which were the crowning for faith. True faith produced generosity and compassionate caring for others including the disadvantaged. These are the steps the Holy Spirit guides each believer on the journey to an abundant entrance to the Eternal Kingdom.

> *1. Since, we stand declared righteous by faith, let us enjoy the possession of peace with God through our Lord Jesus Christ: 2. by Whom we have access by faith into the grace wherein we stand, and we rejoice in the hope of God's glory. 3. And not only, but we also glory in hardships and sufferings: because* **we know that troubles produce patient endurance;** *4.* **and patient endurance approves character; and character brings hope:** *5.* and a faithful trust in God's promises *will never put us to shame; because* **the love which God has for us is poured out in abundance in our hearts by the Holy Spirit which is God's gift.** *(Romans 5:1-5 EDNT)*

This philosophy has permeated my adult life and guided my faith-based outreach, civil rights and all social concerns and operations. Understanding the basic concepts of grafting (*see Grounding p39ff*) may clarify the problem as it relates to the integration of multicultural communities. One does not have to be a horticulturist to see the disadvantage of foliage without fruit or different kinds of fruit on the same tree taking from the root but giving nothing back. Such groups become "takers" without giving a fair share commitment to the infrastructure and knowledge base, which formed the essence and the rootstock for their new reality.

Separate but equal becomes a misleading term and a poor description of **(:=)** *"equal by definition"* without adequate assessment of the situation. This may have gender, marriage, and general family implications, and as the family goes so goes most faith-based operations. Also, as faith-based operations go so goes the nation. This requires deep and prayerful grasp of that which may be strange or wrong. Adjustment and compromise are part of the maturing process and required tolerance even in a faith-based lifestyle.

Some may be in the fight, part of the struggle, but are faint-hearted and unequal in steadfastness. They are blessed with forgiveness of past sins but fail to persevere and endure the hardships of the journey that builds character. All should remember the struggle of early believers who navigated through a maze of Judaism and Greco-Roman mythology and paganism. Yes, they were walking on the way established by Jesus and His Disciples, but there were great trials and tribulations. (see Hebrews 11:32-40)

The US Declaration of Independence secured for all Americans their unalienable rights, *"life, liberty, and the pursuit of happiness."* At the end of the Civil War, President Lincoln used Euclid's Notions to influence the passage of the 13th Amendment to forever abolish slavery as an institution in the USA and territories. The Civil War Amendments (13, 14, and 15) were designed to ensure equality for the emancipated ones and prohibited governments from denying U.S. citizens voting rights based on race, color, or previous servitude, and math had a symbol **[:=]** *"equal by definition"* that opened the door for "equal under the law." However, to say it is true does not make it true! The social change timetable moves slowly.

Regeneration may be seen as instantaneous, but reformation and restructuring a life takes a while! Conversion is only a start, not the finish line. A failure to grow and develop is a serious problem in any aspect of a living system. Life cannot be stopped by a freeze-framed moment; only a corpse is "stopped dead in their tracks!" Never moving beyond the starting place is no way to finish a race. Growth in grace and knowledge is a result of consistent steps in the right direction.

A realization of the change timetable may be seen in the Civil Rights Movement. After Lincoln's Emancipation Proclamation, it took 100 years for Congress to pass a Civil Rights Bill. And individuals who care are still working on the issue. In this process unintended consequences known as **entitlements** have accumulated. Individual rights and privileges became group claims, and an entitled cohort developed who take from the rootstock without giving back. There have been too many entitlement grafts made on the Tree of Liberty. What is the solution?

A century of **Affirmative Action** did not work, what is needed is a personal **affirmative attitude** in the heart and mind of citizens and a willingness to follow the Golden Rule that exists in all major religions. This imperative is based on an **affirmative attitude** *"having* *a favorable predisposition to act fairly, honestly, morally, and without prejudice or purpose of evasion in all human interactions."* Try it! **Affirmative assertiveness works!** You could make a difference in the lives of others and receive major benefits yourself. Considering assisting others, a young girl in St. Croix, USVI said, *"You can't spread jam, not even a little, without getting some on yourself."*

Differences divide and commonalities unite; therefore, the common ground for constructive social change in the global context would be much easier if everyone understood both the process and the problems of grafting. An individual or a group cannot long take from the essence of a dominant culture or the nourishing rootstock of liberty or the spiritual taproot of historic faith-based operations without giving back equal benefits to match their blessings. To balance the equation of Liberty, the beneficiary of benefits must return unreserved loyalty and support for the cause of sharing with neighbors, assisting harmony in institutions, encourage agreement in the workplace, and supporting honest discourse in the public square. It is payback time!

Parents, families and faith-based folk must be caring and compassionate humanitarians to advance human welfare and constructive change for all in their sphere of influence. To receive from a parent or a primary culture's foundational essence requires the recipient to give back in energy and resources equal to what is accepted. All human interaction must have equivalence on both sides of the human equation; all human interaction must be fair and balanced.

This does not mean a "cookie cut operation" of total sameness; even identical twins have individual differences. No fingerprint, footprint, or an eye or ear is exactly the same as another. All differences must be acknowledged, and sameness should be accepted as common ground for advancing understanding. Sufficient sameness indicates common ground; even though there may be a contradiction or some conflicting differences, the presence of camaraderie or cohesion exists among the separate parts of a cohort*. In other words, we can agree to disagree agreeably. This is the "good spot" on the bat for a home run when the bases are loaded.

Integration and assimilation of new believers would be much easier if everyone understood the problems of grafting. Parents, families and faith-based folk must be caring and compassionate humanitarians advancing individual and family welfare and constructive social assimilation for all in their sphere of influence. In most groups or personal conflicts there is a place of agreement or sameness, when this is found genuine progress may be made for constructive social change. (*For more on this issue see Appendix A.*)

*Such as sports, warfare, politics, minority relations

PART THREE: Intermediate Steps

STEP V: MAINTAIN GODLY WORSHIP (∝)
STEP VI: SUSTAIN ENDURING STEADFASTNESS (⚓)
STEP VII: PRACTICE SELF-CONTROL (±)
STEP VIII: LEARN FROM BOOKS AND TEACHERS (:=)

ARE ALL BELIEVERS EQUAL?

THE POWER OF MORAL EXCELLENCE

(α)

Step V

Maintain Godly Worship

(α) *Proportional to something (a stretched alpha)*

And you too have to contribute every effort on your own part, crowning your FAITH with *[1] (benevolent love) and [2] (brotherly kindness) and [3]* **(godly worship)** *and ... (2 Peter 1)*

28. The Kingdom we have inherited is one which cannot be moved; in gratitude for this, **let us worship God as He would have us worship Him, in reverence and godly fear:**
(Hebrews 12:28 EDNT)

    The symbol of a stretched alpha (α) that resembles a fish signifies proportional to something. In mathematics the stretched alpha means that two quantities are related in a linear manner. The fish was an early symbol of the Way. Worship is proportional to godliness. Godly worship includes both believing and behaving. Reciting what one believes is of little value unless the truth of the word is woven into the fabric of faith and results in godly behavior. How can one truly worship God when their heart is not clean, or their mind is burdened with cares? This is why the Word asked believers to *"cast your cares on God."*

    Worship is the outgrowth of lifestyle and is proportioned to *"how much God is valued"* in the daily lives of believers. When believers bow down with a teachable spirit, God lifts them up in a spirit of true worship. In reality, worship is *"to be shut in with God in a secret place."* This shuts out the noise and static of human activity and places one in the spirit of worship. In North America the word *"cocoon"* is used for a *retreat from the stress of public life to the private world of family and friends.* To some degree worship is to be cocooned or "shut in with God in a special place for protection and comfort." An old version **Shut in With God** and the composer long dead and forgotten is worthy of being shared in relation to worship.

Ω

Shut in with God in a Secret Place.

Shut in with God beholding His Face.

Gaining new Power to run in the race.

Oh, I love to be shut in with God!

Ω

Of all places on land or sea,

There's no place sweeter to me,

Then kneeling at the feet of my Lord,

There I'll be shut in with God!

Ω

The pathway though rugged it be,

I'll travel 'til my Savior I'll see,

Then Heaven's gates will open for me,

There I'll be shut in with God!

-Lyrics author unknown

6. Bow down before the strong hand of God, that in His good time He may lift you up. 7. Throw back on Him the burden of your anxiety, because He cares for you. 8. Have a thoughtful demeanor, be on guard, because your enemy the devil, prowls about as an angry lion, seeking someone to greedily consume. 9. Be strong in faith and stand up to the devil, knowing that you share the same suffering with your brothers all over the world. 10. And the God of all grace who has called us to enjoy His eternal glory in Christ Jesus, after you suffer a little He will restore, strengthen, and establish you as a resident of heaven. 11. To Him be glory and power through endless ages, Amen. (1 Peter 5:6-11 EDNT)

The Greeks used a word *kuriakos,* which means *"belonging to the Lord."* The New Testament used this word for *"the Lord's Day"* and for *"belonging to the Lord."* The Greeks had another word *"ekklesia"* used for an assembly of citizens in a city-state. Somehow, *ekklesia* became associated with the early church, probably because the Gentile congregations were filled with Greek citizens, but *kuriakos* "belonging to the Lord or the People of the Lord or the Lords Day"* was a better word to describe a gathering of faith-based believers who much as Abraham, were *"looking for a city with foundations, whose architect and builder was God."* (Hebrews 11:10)

---

*kuriakos is better than *ekklesia* as a foundational construct for the congregation of believers. Perhaps the real reason ekklesia became associated with the church is the General Rules to guide the translators of the KJB. Rule #3 "The Old Ecclesiastical words to be kept, viz, the word **"Church"** not to be translated **"Congregation."** Rome and the Church of England

> saw the church as a "place" not "people." The church is more than a meeting place; it is made up of all true believers wherever they are at any time. The believers are the church, gathered or scattered. After the Reformation based on *"The just shall live by faith,"* some groups called their place of worship, a "Meeting House." In Colonial America the town Meeting House was often used for other functions and gradually congregations began to build buildings and again *"church became a place"* rather than a meeting of those who belonged to the Lord. **Why can't we get back to the pristine formula and worship God in spirit and truth?**

When a prominent person walks into a crowded place no one has to prompt a response; it is automatic. How much more when one feels the Presence of God in sincere worship that others would also respond. Why do we think that true worshippers have to be prompted to respond or entertained to be happy? The "presence" of God should be sufficient to cause churchgoers to show reverence, adulation, and praise to God as they celebrate the value and worth ship of God in their lives. Note Paul's words:

> *24. The God who ordered the universe and all the things in it, the One being Lord of heaven and earth does not dwell in hand made shrines; 25. neither is He served by human hands, as though He needed something from man, seeing He gives to all life, breath, and all things; 26. and has made of one blood all nations of men who dwell on the earth, determined the history of nations and their territory; 27.* **so they should search for God and hopefully find Him although He is not far from all of us. 28. For in Him we live and move, and have our being;** *(Acts 17:24-28a EDNT)*

Sidebar: Worship is from the heart and prompts behavior that glorifies God. During the Vietnam era, I was a reserve USAF Chaplain. Speaking in uniform on a Sunday morning near a Military Base, at the close of the service one could feel the strong Presence of God. A young man in the back quickly stood at attention, braced and stood for a moment then sat. After the benediction, he came quickly to the front and said, **"Sir, I am sorry to interrupt your service. I just spent 4 years in the U.S. Marines and when I felt the presence of a superior officer I automatically stood, braced and waited for orders."** Would it not be a great improvement in worship if the Presence of God were strong enough to be felt by the audience.

This could prompt a spontaneous response by those who need to pray and there would be no reason to end a service with an Invitation. Such a spontaneous response would be the drawing of the Holy Spirit for all to see. Not ending with the Pastor urging the lost to come when the Spirit is not drawing them to God. "When one is convinced against their will, they are of the same opinion still." This can bring a negative end to a time of worship and it has no positive benefit. In fact, the lost should be won at the earliest point in time at the farthest distance from the place of worship. Then converts brought in as believers for spiritual guidance and instruction. Certainly, we do not wish to forbid the lost from entering the House of Prayer, but in Jesus' words, it is not a "den of thieves." If they come, remember salvation is of the Lord and let the Holy Spirit draw them to God. At that point, the believer's responsibility is obvious: converts are "babes in Christ and must be protected, preserved, and nurtured in faith.

General Booth, the founder of the Salvation Army in England had a workable plan to reach the lost. When he was criticized about his approach. Booth responded, "You ring the church bell, and it says **'come to church, come to church'** but they do not come. We go to the streets with a base drum and a trumpet: the drum says **"Fetch 'um, fetch 'um" and we get 'um!"** The trumpet says, **"Jesus saves, Jesus saves"** and they respond to the Good News.

Those who gathered for worship and listening for faith-based guidance were not present to handle community business as citizens of a city-state. In this writer's judgment, the *"ekklesia"* application weakened the concept and construct of godly worship in a House of Prayer for those on a journey to a Heavenly City. The English suffix *"ship"* normally added to personal nouns showed *"condition, character, skill, or office."* In Middle English used during the rendering of the KJB, widely used *"ship"* was for adjectives and participles. Only two of those survive *"**hardship and worship**"* based on the word *"**worthy.**"* Godly worship would then mean some form of worth i.e. value, substance, asset, significance in relation to **"agency, atmosphere, circumstance, competency, spirit or strength."** Look at only one of these: **agency** which means **"He who acts through another acts Himself."** In worship when God acts through a pastor, teacher, or sanctified saint in answer to a fervent prayer through the Word and Spirit it is God acting. Do we recognize the Presence of the Divine? Do we truly value the worth of God in our lives? **This is godly worship!** (SEE APPENDIX B – NAMES AND ATTRIBUTES OF GOD that show His worth and value to believers.)

The early days of *"The just shall live by Faith"* the meeting place was simple, and the people were humble and gathered with a teachable spirit. The atmosphere was quiet but spiritual. Worship began as mind, heart and soul of the first person who arrived for worship. As others gathered, they all waited in silent mediation for the leading of the Holy Spirit. God at times of meditation whispers words of guidance to the ear of the h**ear**t in *"a silent voice."* One had a hymn, another a testimony, some shared a portion of scripture relevant to their life, others witnessed how God was working in their lives, another shared an answered prayer or a special understanding of the Word. An elder would give a scriptural lesson, and prepared believers who were walking in fellowship with the Lord to participate in Holy Communion. There could be prayers for the sick and prayers for loved ones. Godly worship is a gathering of believers who are glad to be a part of the redeemed Family of God. **The worship service is not fellowship, but a place to be shut in with God!**

> *39. Because the promise (of the Spirit) is to you, and your children, and to all those in distant places, even as many as the Lord our God shall call. 40. And with many other words did he testify and exhort, saying, Rescue yourselves from this troublesome generation. 41.* **Those who willingly received his word, were baptized: and the same day about three thousand souls were added to the believers.** *42. <u>And they continued consistently in the apostles' doctrine and fellowship, and in breaking of bread, and in prayers.</u> 43. Everyone was* **filled with a sense of reverence:** *and many signs and wonders were done by the apostles. 44.*

> ***All who believed kept together, and all their possessions were shared;*** *45. Goods and property were sold and distributed as every man had need. 46.* ***And they agreed to meet daily in the temple and to break bread from house to house, and they took meals cheerfully and with personal commitment. 47. Praising God and having favor with all the people. And the Lord added to the church daily those being saved.***
> (Acts 2:39-47 EDNT)

Worship is a vertical experience. Godly worship is universal and all believers regardless of their location who value God in their lives are part of godly worship. Worship is an individual vertical experience that has little to do with who is present or absent. One needs to feel *"shut in with God in a secret space"* to truly worship. Regardless of the venue or a solitary place of prayer, worship is an individual matter; it is not group fellowship. Worship (^) is vertical, and fellowship (<>) is the horizontal part of faith-based behavior but is not considered worship. Fellowship is a time when believers value each other: pray for one another and for those unable to participate. Fellowship may be part of a time of celebration of spiritual matters, but one should always have a demarcation boundary to differentiate fellowship and worship which clearly separates the time of gathered believers for godly worship from other meetings where the program is fellowship, spiritual celebration, guidance, joyful music, or preaching. Such meetings have value for the collective body of believers, but "worship is the celebration by individuals of the worth and value of God in their personal lives. Fellowship is a group function, but worship is an individual response to the worth and value of God. Just as salvation is a private and personal matter, worship is a ***"me time with God"*** where distractions are shut out and the individual is shut in with God Presence.

Of course, there are other separated saints who may join in the spirit of worship at the same time, but it is a personal interaction between an individual and God. Only the people who truly *"belong to God"* can achieve godly worship. When several individuals have a *"come to Jesus moment"* in the same place at the same service, this is not group conversion. It is multi-individual decision making. This holds true when a congregation has a simultaneous response in worship because the worth of God has been woven into the fabric of their faith. Such a response is often called celebration when in reality it is individual worship. Others could be caught up in an artificial response to the excitement of the moment and enjoy the service without valuing the worth of God in their lives. This is where a line must be drawn between entertainment and worship. A good hymn may produce godly worship provided the words remind one of the worth and value of God. The Pastoral Sunday homily that has a teaching element that identifies weakness and failures in the congregation may need to be "endured" instead of "enjoyed." Godly worship is a personal matter {not to be drummed up} because godly worship is an automatic response to the value of God.

> *1. I charge you before God, and the Lord Jesus Christ, being about to judge the living and the dead at the appearing of His kingdom; 2. Proclaim the word, immediately and at inconvenient times; warn, reprimand, urgently encourage with all long-suffering and teaching.* ***3. There will come a time when men will not tolerate healthy teaching;***

but following their own desires shall listen to many teachers because they are impatient to hear something to please and gratify their ears; 4. and they will stop listening to the truth and be turned aside to fictional tales. 5. *You must be clear-headed in all things, endure hardships, declare the good news, fulfill your ministry. (2 Timothy 4:1-8 EDNT)*

Perhaps a study of the God-ordered Temple could shed light on God's perspective of worship. Basically, there were the (1) Stairs of Assent to Double Gates, (2) the outer court where people assembled, (3) an outer chamber called the sanctuary or Holy Place, and (4) the Holy of Holies where only the highest of spiritual *leadership* were permitted to enter on behalf of others. There were dangers when entered improperly or without a right relationship with God. If one entered without God's anointing with sin in their lives, God struck them dead. The spiritual leaders who entered had bells and pomegranate sewed to their garments so the people could hear the bell and know that God had accepted their prayers, their worship and their sacrifice. The rattle of dried seeds in the pomegranate was evidence of past harvest and seeds for future harvests. The bell signified the Presence of the Spirit of God and the rattle of the pomegranate seeds was evidence of past harvest with seeds for the future. Genesis 1:29 was clear *"The seed is in the fruit"* for future harvest. Also, there were calendar dates and time slot limitations. Why? Because the worship of the One God, the Creator and Sustainer of the Universe in a sanctuary set apart for worship, reverence, respect, adulation, praise and admiration of the worth and value of God in all things.

The gathering for worship was not protection for evildoers, but a place of forgiveness, reconciliation and restoration of hearth and home. It was not designed as a place for shelter only, or just for physical food and protection, but a gathering of family and friends; it was not a safe haven for thieves, bandits, or those who did not appreciate the sacrifice for sin in exchange for Eternal Life for those who accepted God's grace and forgiveness.

God's house of prayer was respected as a sanctuary for believers, not a social club or restful retreat for thieves. Certain parts of worship should only be entered by those who have cleared their conscience through confession and are clean before God and worthy of Holy Communion. In fact, there are serious physical and eternal consequences for not doing proper self-examination before participating in worship or taking Holy Communion in an unworthy manner. The true purpose of special sacraments and procedures of worship are not fully understood. This unworthiness included: *not properly considering the Lord's Assembly.* There must be proper respect and honor given to the procedures of worship and the venue which is to be a *"Gathering of Believers who belong to the Lord."*

> **27. And in conclusion, whoever shall eat this bread and drink this cup of the Lord, in an unworthy manner, shall be guilty of violating the body and blood of the Lord. 28. Each time let a man examine himself before he eats of that bread and drinks of that cup. 29. The reason for self-examination before eating and drinking is to prevent partaking in an unworthy manner when one does not properly consider the Lord's assembly. 30. For this reason many are**

powerless and sick among you and many die.
31. For if we judge ourselves rightly, we should not be judged by God. 32. But when we are judged, we are disciplined of the Lord that we should not be condemned eternally with the world. 33. Wherefore, my brethren, when you come together to eat, wait in turn for a proper distribution. 34. If any man is hungry, let him eat at home that there be no confusion when you come together to eat.
(I Corinthians 11:27-34 EDNT)

Worship may mean different things to different people; however, the essence of worship is a realization of the *"worth-ship of God"* and is a vertical (^) experience. Worship has little reference to those present; it is a concentrated exposure to the Presence of God through prayer, song, witness, listening, believing and behaving God's Word. Any distraction can be a diversion from worship. How much is God worth in your life and your family? How do you value God? Worship is a serious and spiritual observance of the redemptive value and personal sacrifice brought to believers by the death and resurrection of Jesus. Godly worship is looking at the Cross through the empty tomb and realizing that Jesus is alive and His Spirit is always with us on our journey. Worship is respect for the holy (respect is to observe and pay attention to) that which has value in life. One may worship, through intercessory prayer, by studying, listening, learning, believing and behaving the Word. Preceded by a moral lifestyle, confession, and a clear conscience before God opens the windows of Heaven for the blessings of godly worship.

Churchgoers internalize their needs and personalize their individual approach to the worship of God and resist any intrusion from others, especially strangers. The personal interactive intricacies and restraints at work in spiritual worship become a freeze-frame view similar to one person watching a televised worship service in the privacy of their home. Some desire to be *"shut in with God in a secret place."* When like-minded individuals gather together to personally celebrate their salvation and the worth of God in their life, these kindred-minded souls enjoy association with others through a kinship-type of response normally known as *camaraderie*, which is mutual trust and friendship among people who spend a lot of time together. It is a feeling of pride, fellowship and/or shared loyalty among friends. This is interaction at the human level and is one step below the personal worship of God. We should remember that conversion is a personal interaction between a contrite heart and God. True worship is also personal and requires treasured attention to the words of songs and the Word of God without the distraction or intrusion on their mind or personal space by intruding* strangers. Worshipers remain individuals. Congregation is a gathering of like-minded souls; however, worship is not fellowship (<>) or a group activity. Several may worship God at the same time, but it is still individual and personal response to the Presence of God.

>*To put oneself deliberately into a place or situtation where one is unwelcome or uninvited (see p117)

Worship may be an individual response to the "worth-ship of God" or a collective expression by 2 or 3 or by hundreds or thousands: there are no limits to the number included in the experience of godly worship (one to infinity). *"Where two or three are **gathered together,*** *there am I in the midst."* The emphasis is "togetherness" not numbers. It seems to be the concept of consensus or one accord; the Greeks called this *homothumadon,* or *"moving along in harmony and togetherness."* Worship itself becomes an observance or remembrance of the common gifts of faith and shared promises of God to people of faith. Entertainment, personal excitement or an emotional response are not true expressions of worship. At times God ministers to believers in a *"still small voice."* Godly worship provides personal blessing and has lasting results. When the Spirit of worship is present some will feel condemned and repent, others will feel content and acknowledge the value and worth of God in their lives. This is a good thing; it enables some to remain current in their relationship with God. True worship is a preparation for Holy Communion.

> *24. The God who ordered the universe and all the things in it, the One being Lord of heaven and earth does not dwell in hand made shrines; 25. neither is He served by human hands, as though He needed something from man, seeing He gives to all life, breath, and all things; 26. and has made of one blood all nations of men who dwell on the earth, determined the history of nations and their territory;* ***27. so they should search for God and hopefully find Him although He is not far from all of us. 28. For in Him we live and move, and have our being;*** *(Acts 17:24-28a EDNT)*
>
>> **Sidebar:** Often traveling alone on trips, my elderly mother would occasionally travel with me to "see the country." On one trip to Washington, DC we ventured off the Interstate to see the small towns in Virginia. Mother observed the number of eating places and the quantity of church buildings and said "If anybody starves to death and goes to hell it is their

fault. There are places to eat and worship on every corner."

Worship should lead to the outreach of the message of grace. Believers were saved to serve, and worship is part of God's renewal process to keep people of faith current in their standing with God. Entertainment, personal excitement or an emotional response to songs or sermons are not true expressions of worship. At times God ministers to believers by searching the inner-most part of the soul for thoughts that could lead them astray from the right path. God's benevolent love has constant concern for the spiritual health of believers. This tender loving care enhances the attitude which produces godly worship and has lasting results expressed in witness to others.

> Search me O God, and know my heart,
> And test my thoughts.
> Point out anything in me that makes you sad.
> Lead me along the path of everlasting life.
>
> (Psalm 139:23 TLB)

It has been my custom to purchase a new Bible every seven years and with the first sermon, normally, the new Bible is initiated with a statement that is written in the front of the book. Speaking at Kensington Temple in London, UK the first message from a new Bible, this inscription was shared: **"When an unconverted person reaches the sanctuary of the church unsaved, it is evidence that the local plan of evangelism is not working."** True evangelism is believers witnessing to the lost at the earliest point in time at the farthest distance from the sanctuary as possible. All mature believers must wear the full armor of God and have "swift feet" to take the message of grace to the community.

> *14. How shall they call on Him in whom they have not learned to believe? And how shall they believe in Him of whom they have never heard? And how shall they hear without a messenger? 15. And how shall they proclaim, except they be sent. As it is written, Fully, developed are the swift feet of those who proclaim the glad tidings of the gospel!* (Romans 10:14-15 EDNT)

Then when one has learned to believe, they are brought to the House of Prayer for guidance in spiritual growth and development. The present system of asking individuals to come forward and deal with their sins before God in public is not found in Scripture. Could this be why so few new converts are counted? Does the Word not provide a better way? Jesus is quoted in John 6:44 "No one comes to me except the Father draws them to me!"

Godly worship of gathered believers shows the "worth-ship of God in their lives." It is not a meeting of civil citizens who have no spiritual connection to the Creator. In times past the congregation included families and friends, but now faith-based meetings are a gathering of strangers without a redemptive connection to the divine. And the worship is mostly entertainment meant to satisfy the whim of the people rather than gospel hymns that communicate the faith-based message. The congregation has become an artificial dichotomy of spectators and participators with irreconcilable differences. The unity of the Spirit and the bond of peace are nowhere to be found. The false gossipers are there, the hypocrites and unashamed backsliders whose name is on the church roll, but may not be in the Book of Life, are there, and mixed in with this mess are a few sanctified saints who desire to truly worship God. This is no way to do God's Work God's Way!

Worship may include confession, collect, communion and celebration. *Confession* is an individual act of admission of human weakness, the acknowledgement of guilt, and the affirmation of forgiveness. *Collect* is the gathering "together" of believers assembled in unity responding to personal pardon and the promises of God; this is personal prayer, praise and worship. *Communion* is born again believers walking in fellowship with God and others commemorating the death and resurrection of Jesus; it included what the Greeks called *"koinonia"* and transliterated as *communion, community, participation, stewardship, sharing, and spiritual intimacy.*

The first use of **koinonia** was in Peter's sermon at Pentecost (Acts 2:41-47) where the essence of *koinonia was love, faith, and encouragement.* Also, Paul used the concept to express agreement with one another, being united in purpose, and serving alongside each other based on fellowship with God and each other. (Philippians 2:1-11). Communion clearly expresses that each participant believes they are current in their relationship with God and others. Communion is true *koinonia* and suggests shared beliefs, common expectations, closeness of relationship and remembrance of the death, burial and resurrection of Jesus.

Gathered believers have been called God's Farm or God's Garden where good things grow. Now many gatherings have the atmosphere of a graveyard where the occupants are dead, and the flowers are dying or dried up. What does it take to grow a garden? Diligent cultivation, careful prayerful planting, constant attention, unceasing effort and saintly patience.... then it must be touched by the Hand of God. Early spiritual leaders were cultivators of soil, seed sowers, planters, vineyard keepers, fruit pickers, and harvest gatherers provided their labor was blessed by a touch of the Hand of God.

> **Sidebar:** My Grandfather Green had a prize-winning crop of corn. One Sunday a city slicker visited his church and was invited to Sunday dinner. At the table Grandfather mentioned his prize-winning crop of corn. The city slicker (who knew nothing about raising corn) said "You should give God more credit for your crop. He furnished the ground, the sunshine, the rain and the minerals in the ground." The response was clear, "You should have seen that field when God had it by Himself. He sure left a lot for me and the boys to do!"

GROWING A VEGETABLE GARDEN

DILIGENT CULTIVATION
CAREFUL PLANTING
UNCEASING EFFORT
CONSTANT ATTENTION
SAINTLY PATIENCE
...THEN IT MUST BE TOUCHED BY THE HAND OF GOD!

A spiritual assembly is a growing community of believers committed to accelerating the multiplication of healthy, reproducing converts, disciple making, teaching, fellowship and worship that can grow into additional local congregations or mature into a new and established place of worship and further expand the kingdom of God.

God planted a garden and placed man there to care for His handiwork. Paul called believers God's garden or field to be worked. Growing a faith-based entity is much the same as working a vegetable garden: **Why can't good people see that connection?**

Just as plants in a garden must be spaced properly or there would be cross-pollination and the crop could be spoiled., the space immediately surrounding someone is known as "personal space." Any encroachment on this area causes one to feel threatened or uncomfortable. This space is defined as "intimate" about 18 inches and "personal" from 1½ to 4 feet where only friends and acquaintances are welcome. Strangers are strictly forbidden in the intimate and personal space. Then, there is "social space" which extends from 4 to 12 feet where people feel comfortable interacting with acquaintances and friends. There was a time when a morning worship service was a gathering of family and friends, but now in a pluralistic society, faith-based gatherings include the curious, the unconverted, carnal but professing believers, strangers, visitors, and a few individuals that are being drawn by God toward repentance. It is easy to see that the rules of "personal space" are at work in worship. Most women prefer not to be hugged by strange men in church. Some men cannot handle a beautiful young lady within their intimate space. This reminds me of an event in my early ministry.

> **Sidebar:** After ministering in a Jacksonville, FL Church, standing with the pastor, H. G. Poitier, at the rear of the black church, several beautiful young ladies in the line began to hug the old pastor. He was a handsome old gentleman, Sidney Poitier's uncle, and had been their pastor for many years. They had

grown up under his ministry, so they gave him a full bosom grandfather hug.

As they came to me, stepping back with an extended hand seemed appropriate. Some reluctantly shook my hand, but one beautiful black lady insisted on hugging me. My response, *"I'm not sure my ordination can handle such a hug!"* Her reaction was transparent: *"You white folks make sin out of everything!"* My notion that young ministers should err on the side of caution troubled her. She mumbled something like, *"You white folks can't tell the difference between Christian love and that other stuff."* All the old preachers counseled me to always be cautious about over familiarity with females. My thoughts were **"...it's better to be safe than sorry."**

Nothing should be allowed to distract individuals from godly worship. The presence of the unconverted or the immature during worship may well be a distraction and cause a cloud to overshadow the congregation. Worship is a private and personal time with God when all distraction is shut out and one feels alone in the Presence of God. The old Hymn expresses this state of grace:

> Shut in with God in a secret place.
>
> Shut in with God, beholding His Face.
>
> Gaining new Power to run in the race,
>
> Oh I love to be shut in with God.
>
> —Unknown

Step VI

Sustain Enduring Steadfastness

(⚓) *Symbol for security, safety and steadfastness*

And you too have to contribute every effort on your own part, crowning your FAITH with [1] (benevolent love) and [2] (brotherly kindness) and [3] (godly worship) and [4] (enduring steadfastness) and ...

In the days of sailing ships, an anchor was a great safety tool to secure a vessel from drifting in the wind or with the current. They were especially needed in a storm and when anchored in harbor. **Believers need a safe and sure anchor in troubled times to provide hope and enduring steadfastness.** Hope is contingent on two things: 1) **desire,** one must want something to happen, and 2) **expectancy** (faith) they must also believe it will happen. Without both, fear develops that it will not happen. The anchor was an early symbol of hope and steadfastness for believers.

> *...we who have fled to Him for refuge, might be strongly encouraged to lay hold upon the hope that is set before us.* ***19. This hope to us is an anchor of the soul, safe and sure, and it enters with us to the inner court beyond the veil.*** *(Hebrews 6:18c-19 EDNT)*

Family members often permit a child to be "little adults" by omitting basic steps in the growing process. This is probably the greatest hindrance to mature development and spiritual formation. Normally, one is taught the essential elements of a subject before moving to more complex subject matter. One does not climb a mountain starting at the top, or climb a ladder starting in the middle. The first steps must be carefully negotiated. Human nature and nurturing of the young require early faith-based steps for foundational development. Children are not taught to add a column of figures starting in the middle or at the top. Starting with the foundation stones and move step by step upward is the normal process.

An early lesson in arithmetic is the check and balance process. One adds to get an answer then subtracts to confirm; the same process is true with division first one divides then multiplies to confirm. In mathematics there is always a balance or equality on both sides of the equation. In faith-based living there must be balance between knowing and understanding and/or believing and behaving.

> **Sidebar:** Note an example in Luke 10:25-37. A certain man had memorized spiritual material without understanding or learning to believe and behave what had been learned. This made the process impossible to complete because he did not recognize his responsibility toward others. **Yes, he memorized scripture, but it was not woven into the fabric of his faith.** He was preoccupied with desire for eternal life instead of understanding that he must believe and behave the Word and understand the process of reaching moral excellence. Someone had failed to complete the task of preparing him to make a **constructive** decision when given the opportunity. There was no positive decision to launch him on a plan to construct a faith-based lifestyle.
>
> *25. A certain lawyer stood up and tested Jesus, saying,* **Teacher, what shall I do to inherit eternal**

*life? 26. He answered, What, is written in the text? How do you read it? 27. He answered, you shall love the Lord your God continually with your **whole heart**, and with your **whole soul,** and with your **whole strength,** and with your **whole mind;** and **your neighbor as your own self.** 28. Jesus said, correct do this and you shall live. 29. And he, willing to justify himself, said to Jesus, and who is my neighbor?* (Luke 10:25-29 EDNT)

The emphasis by position based on Greek syntax must be noted above. The most important is at the bottom of the ladder and the goal is at the top with logical steps between. The form of *agape* used here has a forward look. The order starts at the base, "love yourself" and moves incrementally step by step upward "Love God continually with your whole heart." Each step is crucial to the next one and the ultimate objective of loving God continually with the whole heart. One may not start at the top of the ladder. The first step is to understand that God does not make junk: you must love yourself before you can truly love God and others. Note the emphatic order of the following (1-7) steps:

7. Love God continually
   (adamantly, constantly, steadfastly)
6. Love God with your whole heart (core, being, focus)
5. Love God with your whole soul (mind, will, emotions)
4. Love God with all your strength
   (physical, Intensity, energy,)
3. Love God with all your mind
   (knowledge from books and teachers)
2. Love your neighbor as yourself
   (love those near you)
1. Love yourself (God doesn't make junk)

How does this happen? God forgives past sins for all who come to Him with a remorseful heart and seeks forgiveness. With this divine absolution comes a kick-start in the direction of constructive personal change. Each individual has a "measure of faith" to use wisely on their journey. Once it is understood that this faith must be used every day, believers are required to utilize their faith by adding benevolent love, then brotherly kindness, followed by godly worship with a steadfast attitude adjustment and self-control. Then to be fully approved by God, believers must learn from books and teachers and ultimately crown their faith with moral excellence expressed in lifestyle. When this process is working, believers will be productive in the experiential knowledge of the Messiah's Lordship.

> *10. So, believers, be the more eager to confirm your calling and your choice: for if you do practice these virtues, you will make no false steps: 11. and you shall be richly supplied the entrance into the kingdom... (2 Peter 1:10-11)*

> *8. Now God is continually able to overflow you with self-sufficiency always making you competent to pour out to the good of others: 9. as it is written, His generosity is scattered to the poor; His love-deeds are never forgotten. (2 Corinthians 9:8-9 EDNT)*

> *20. And they went out and witnessed everywhere, the Lord working with them, and validating the message with accompanying supernatural wonders. (Mark 16:20 EDNT)*

Worship embraces the value of God and acknowledges the worth-ship of divine influence in all aspects of life, for the individual and collectively for those gathered. Worshipful celebration is an expression of *agape* love (one-way love that is vertical and spiritually direct).

> **Sidebar:** In the old West a cowboy attended a Brush Arbor Revival and responded to the minister's invitation to surrender to God. The next Friday night as he came back to town, his horse trotted up to the saloon and the cowboy jumped off, tied his horse, and went smiling through the swinging doors. A loud cry came "John, thought you got religion last week!" Embarrassed the cowboy said, "But my horse wasn't in on the deal. Looks like I have some horse training or trading to do!"

> .... *Rescue yourselves from this troublesome generation. 41.* **Those who willingly received His Word, were baptized:** <u>and the same day about three thousand souls were added to the believers. 42. And</u> **they continued consistently in the apostles' doctrine and fellowship, and in breaking of bread, and in prayers.** *43. Everyone was filled with a sense of reverence: and many signs and wonders were done by the apostles. 44. All who believed kept together,* (Acts 2:40-44a EDNT)

> *1. If there be any* **encouragement** *in Christ, if any* **reassurance in love,** *if any* **participation of the Spirit,** *if any* **tenderness and compassion,** *2. fill up my joy by* **living in harmony, having the same love, being in one accord of one mind.** *3. Let nothing be done through argument or excessive pride; but in true humility let each value others more than themselves.* **4. Look not after your own interests, but practice looking after the interest of others.** *(Philippians 2:1-4 EDNT)*

Social change may require the mixing of community traditions and cultures. By adding some aspect of one culture to another, the assimilation produces common ground for change. The combination of traditions produces transformation and adjustments to both thought and process. As change policy is developed answers are worked out in advance for anticipated questions and a positive attitude for interaction becomes a dynamic policy for action. This tactic produces a strategy for constructive change as individuals adjust to the differences around them. The ability to accommodate and adjust to a new environment or a larger association is essential to maturity and valued behavior.

There are as many approaches to behavioral change theory as there are individuals; in other words, change has a personal element. As individuals function within a given culture or tradition, they constantly seek to maintain the major aspects of their personal culture but are willing to accept incremental steps toward both intellectual and behavioral change. There is a human yearning to reach for a better way of life and an ambitious striving toward higher goals. This human attribute is an asset to faith-based change. Of course, there are restraining forces that obstruct such progress. The restraints that prevent progress come in the forms of persons, traditions, and certain aspects of culture: food, clothing, music, religion, politics, and personal and social distance. As one makes a mature decision to begin a faith-based journey, they become willing to overlook traditional and cultural warnings and at times negotiate both social and personal distance to better understand new acquaintances. It is this course of action that facilitates constructive transition to a faith-based lifestyle.

In the process of faith-based living there is concern that one does not become tainted by the moral deficiencies of another cultural tradition or individual behavior. It is in this regard that one must be vigilant, willing to take a cautious look at other cultures and traditions, but to also be discerning and accept only those aspects that do not violate their moral standard. The intention is to allow good to overcome evil rather than to sanction immoral behavior. Should one through human weakness accept the dishonorable aspects of life and attempt to imitate shameful behavior, a progressive debauchery establishes a slippery slope toward evil that produces deterioration and decline in moral values. In fact, this is what makes the world go around or at least move forward toward common ground known as social progress.

All citizens of a civil society must be on guard against moral decline that ultimately blocks all constructive progress. The process of integration however is designed to make whole or new by adding or bringing together different parts. The study of theology and/or philosophy creates one's value system and ideology. At the level of ideology and values, different individuals and divergent groups find common ground to effect cooperation. This takes place as a formation in the affective domain where ideas of an individual or class are derived exclusively through feelings. Since feelings can be deceptive, the affective domain must be balanced with a basic philosophy through a study of the processes governing thought and conduct including aesthetics, ethics, logic, metaphysics, morals, character and behavior. This combined with a morsel of theology that considers the relationship between the Divine and the human race as to matters of faith and behavior can open the door to a faith-based lifestyle.

At the death of Martin Luther King, Jr. others found a list on his desk of 100 men most interested in Civil Rights in the South. My name was on that list. As a clergy/educator, it pleased me that someone of another race had recognized my interest in equality and justice. It was firmly believed that social policy could be affected without political upheaval; that real progress could be made in integration without the rancor and animosity of politics. King's non-violent approach was of interest to me. God is not the Author of confusion, injustice, or discrimination. There is no partiality with God! God made of one blood all nations and this human equation must be balanced by faith-based people.

(±)
STEP VII

Practice Self-control

(±) *Plus / minus used for range*

5. And you too have to contribute every effort on your own part, crowning your FAITH with [1] *(benevolent love)* and [2] *(brotherly kindness)* and [3] *(godly worship)* and [4] *(enduring steadfastness)* and **[5] (self-control)** and ... *(2 Peter 1: 5ff)*

8. Finally, you must think the same thoughts, share difficulties with one another, having automatic interdependence with brotherly kindness; be tender-hearted and humble-minded: 9. you must not repay injury with injury, or hard words with hard words, but bless those who curse you. **For you were called to give kind words to others and come to a well-spoken eulogy at the end.** 10. For the one wishing to love life and see prosperous days, let him avoid an evil tongue and cunning words. 11. Habitually avoid evil and do good; let him seek and follow peace. *(1 Peter 3:8-11 EDNT)*

Self-control (±) is the ability to control oneself and requires both plus/minus involving three components: (1) **compulsion**- the pressure to make quick decisions without considering the unintended consequences; (2) **emotional reactions** –thoughtless responses in sensitive situations; (3) **eager aspirations** – materialistic, self-improving desires and yearnings which may be disapproved by others. This requires the restraint aspects of self-control which exercises influence over feelings, emotions, and personal desires. Self-control demands perseverance, positive discipline, determination, stamina, strength, and willpower. Some believe self-control is determined by genetics, but most understand it can be learned and strengthened. Self-control is an achievement, but also is fragile and can be weakened by circumstances and even depleted by extended selfish behavior or a failure to manage anger. Righteous anger against sinfulness is normal but must be controlled and handled with a teachable spirit.

> *25. Wherefore speak every man truth with his neighbor without falsehood: for we are members bound one to another. 26.* **Have righteous anger without sin: let not the sun go down on your anger:** *27. Neither give an opportunity to the devil.* (Ephesians 4:25-27 EDNT)

> *25. And everyone who enters the race* **practices rigid self-control.** *They do this to win a wreath that will soon wither,* **but we seek a crown that will not fade.** *26. I run but not aimlessly; so I fight, but not as a shadow boxer: 27. but I beat my body black and blue and bring it into subjection: lest by any means, when I have preached to others, I myself should be rejected as a worthless coin.* (I Corinthians 9:25-27 EDNT)

Normally, one does not start a career by hanging his hat in the CEO'S Office or asking for a key to the Executive Washroom. The young are not little adults and older folk are more than grown-up children. Age specific lessons and guidance are required. One may expect too much of both the young and the old. Patience which is positive self-control are required for both ends of the age spectrum.

Adversity often becomes an opportunity for achievement. Self-control is the virtue that enables one to know when to act, and when not to act. Self-control is the exercise of patience, the ability to remain under the pressure of trying circumstances until a way forward is found. Patience is the mature capacity to accept or tolerate delay, trouble, or suffering without getting angry. Epictetus, a Greek philosopher, said, *"It is not what happens, but how it is handled that matters!"* Patience and self-control show quality and courage of an individual.

A saying attributed to Napoleon about the hazardous struggle of battle is appropriate here, *"There is a time in each battle when both sides have lost. Victory comes to the force which attacks after this point."* In philosophy, a positive implies a negative; however, the converse is true but can only be constructed by the process of elimination. If one says, "I am going to Atlanta" others know where that person is not going. However, if the statement were "I am not going to Atlanta." No one knows where this person is going. In such a quandary one should reason by following Ockham's razor *"The simplest solution is usually the best and should be tried first."*

A story from early television tells of a young mother who asked an Interviewer *"May I say something to my young son at home?"* Given permission, she said *"Billy, whatever you are doing, stop it!"* It is natural for all of us to do things that displease others. The solution: "stop it!" In sacred writings we find weak and failing individuals selected for a great work and this despite their shortcomings. God often selects the young to confound the wise or the weak to overcome the strong.

When one is divinely called for a purpose, qualification is not the issue; it is availability that counts. There is always enablement for those chosen to serve. Even the faint of heart are used to accomplish great things for the good of mankind. Through the power of forgiveness and divine anointing, individuals are enabled to make positive achievements because God qualifies whom He calls. The work of God in one's life, however, does not excuse the human contribution to increase faith and maintain self-control in the midst of a trying situation. Neither does it exempt one from seeking relevant knowledge from books and teachers. As a Military Chaplain during the Vietnam era, it was shocking to see a young man who a few days before could not make up a bed, find his socks, or carry out the garbage, become an honored war hero. Arising to the urgent need to save others was usually the impetus. The courage of a war hero is the ability to overcome fear facing deadly danger.

"Greater love has no man than a man lay down his life for his friends." (John 15:13)

The sacred record is filled with individuals who failed from a lack of self-control at some point in their life and disappointed themselves and others by their behavior. Adam and Eve made a big blunder in the Garden, but God used them to procreate the human-race. Noah got drunk, but God trusted him to build the Ark that saved the seed family of humanity. Abraham lied about his wife, but God used him to build a Nation. Jonah ran from God, but he was given a divine message for Nineveh. Moses was slow of speech, a stutterer, and had a bad temper, but God gave him the Law and used him to lead Israel out of bondage and servitude. David, the Shepherd

Boy, felt unqualified, but God made him King; he then misused his position and power to take another man's life and wife. Surely, there were negative consequences to these failures, but David was remorseful and asked for forgiveness and he was used as *"a man after God's Own Heart!"* More recently, Mark deserted the missionary trip with Paul, and Mark was refused when he desired to go on another spiritual journey. Later Mark made a drastic recovery and at the end of Paul's life, he desired to have Mark visit him saying *"He is profitable for my ministry."* The positive aspects of overcoming a difficulty provides strength for future trials. The loss of self-control is a tragedy, but maintaining self-control is an achievement that is well rewarded when observed.

> **Sidebar:** There is a story out of Africa where a tribe initiates a young boy into manhood. It seems they send him alone into the jungle to find his own food, protect himself from wild animals, and survive a number of days and nights and return a stronger, braver, more courageous young man. He was taught that each hardship or difficulty he overcame would strengthen his mind, body and spirit. At first, he is a frightened young boy, then he encounters a difficulty on the path and overcomes. Then he meets a small animal and kills it for food; then another larger more vicious animal and slays it. Each time he overcame a circumstance or a dangerous animal, his strength and courage were increased substantially. By the time his trial period ended, he was no longer a frightened little boy but had the strength and courage of each difficulty he had overcome. He was now ready to take on adult responsibilities. He was no longer a liability to the village; now he had become an asset and was ready and able to take his rightful place in the tribe.

Gideon was insecure and wanted a large army, but God limited him to a few good men (so God could be credited with the victories). Thomas was a doubter, but he saw and believed. Peter cursed, lied, and denied the Messiah, but God used him to strengthen others and lead the pristine group of believers. Paul persecuted believers and was struck down on the road to Damascus, but God used him as an Apostle to the Gentiles and trusted him to write one-fourth of the New Testament. Zacchaeus overtaxed the poor but invited Jesus to his house and made restitution. Zacharias and Elizabeth stopped praying for a son, but an Angel said, *"The prayer you no longer pray, God heard!"* (Luke 1:13 EDNT) God enabled them to become the elderly parents of John the Baptist who prepared the way for the Messiah. God does not always call the qualified, but He always qualifies the called. God desires that believers live a life that is faithful to Him, and one that controls all selfishness in relationship to others. Ministry and service to others is based on availability and attitude not ability and qualification. God uses those who are willing to believe, obey, and behave the Word.

It is clear from life experience that adversity can be overcome with patience and self-control. Most of us have experienced times when there seemed to be no way forward. However, adversity became an opportunity to learn to plow around the stumps and move ahead with pressing matters *(see the sidebar below)*. Also, failure is not the end, but a lesson learned and a starting place for better things. Of course, lasting achievements do not come without hard work, but hardship and misfortune require self-control and prepares the mind, body, and spirit to overcome future difficulties.

> **Sidebar:** A pioneer family traveling with a wagon train became weary of the journey and decided to stop and build a cabin. The men cut trees to clear the land, trimmed the branches for firewood and measured and cut the logs to build a cabin. Then they "plowed around the stumps" to plant a garden for future food and took a break and hunted for meat to sustain the family. Now, they are back to building the cabin before winter.
>
> In the spring the wife fell over a tree stump that almost blocked the front door and complained, "Why didn't you dig out that stump when you were building the cabin?" The Pioneer said calmly, "Had I taken time to remove the stump, there would be no food and no cabin. The winter would have overtaken us without shelter, food or firewood. We had to plow around the stumps to make a garden." After the stump began to decay, one evening he placed a log chain around the stump and with a team of horses pulled the stump and roots out. Now everyone was happy, but it took patience and self-control to wait for the proper time to remove the stump from the entrance to the cabin. Hopefully, a lesson was learned and stored in the memory of the children.

Are All Believers Equal?

The Power Of Moral Excellence

(:=)
STEP VIII

Learn from Books and Teachers

(:=) Equal by Definition

5. *And you too have to contribute every effort on your own part, crowning your FAITH with* [1] *(benevolent love) and* [2] *(brotherly kindness) and* [3] *(godly worship) and* [4] *(enduring steadfastness and* [5] *(self-control) and* [6] (knowledge from books and teachers) *and ...* (2 Peter 1:5-7 EDNT)

A function that represents the distribution of many random variables as a symmetrical curve is considered normal; however, a distribution may have a negative or positive skew to the left or to the right as shown in the cell below: A basic understanding of distribution may improve the awareness of those in positions of leadership and increase the grasp of participation in learning from books and teachers.

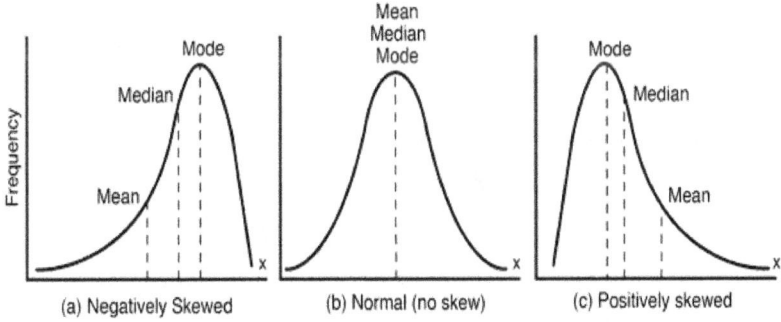

Wikimedia.org: By Diva Jain

Distributed (Negative, Normal, or Positive)

Luke shared the words of Jesus: *"Surely the blind cannot guide the blind or both will fall into a ditch?* ***40. The learner is not above his teacher: but everyone who is fully taught will reach his teacher's level."*** (Luke 6:39-40 EDNT) The key here is "fully taught." Sharing subject matter alone is not teaching, in addition to a knowledge base upon which to build, all learning requires preparation, participation and partitioning.
(1) Preparation to be adequately prepared to receive subject matter transfer. **(2) Participation** as an informed individual on the subject matter to be directly involved in the learning process. **(3) Partitioning** finding ways to handle complicated data by dividing and associating facts by course objectives to increase understanding and learning of the basic aspects of the subject.

The ability of the presenter to *"stimulate interest and arouse a spirit of inquiry"* becomes a factor in the learning process. Prior preparation of all concerned counts, but human nature normally takes the way of least resistance. As a result, the participants will vary in the "value added" quality of their participation. There must be an intersection of both intent and interest in the subject at hand and a connection between (teacher—content—learner) to make the teaching/learning process sufficiently transactional to work effectively.

Many fail to realize that prior study and focused participation determine the level of learning for all present. Some are participators and others are spectators. The reader only understands part of a book; the student only learns part of the subject matter presented by a teacher, a teacher can only teach what they know, and churchgoers only receive a small part of a Sunday homily that is relevant to their lives. These are some of the limitations in the process of subject matter sharing regardless of the venue or the content.

Regardless of negative feelings about mathematics, the distribution function is operative in any group or gathering. Note the normal percentage (%) distribution of the whole:

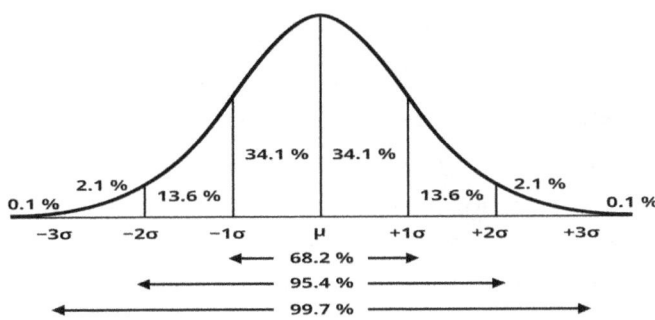

Normally groups or gatherings are divided into degrees of participation and degrees of onlookers. Whether a public classroom, a study group, a political gathering or a faith-based meeting. For example, Jesus chose Twelve (12) Disciples from among His followers as learners to become His messengers. Limited data available complicates a placement of individual Disciples into a distribution model; however, it appears that a majority were average fellows, yet four (4) were His closest and trusted friends: James, John, Peter and Judas had a special connection with Jesus . Although, Judas and Peter had trusted positions: Judas handled the money for the group and Peter served as Sergeant at Arms (he carried the only sword in the group). Provided the Twelve were assessed based on attitude, disposition, and participation, they may appear (:=) equal by definition (they were all disciples chosen by Jesus) but there would be both common human weakness and different levels in prior knowledge and the current state of learning among the group. Notwithstanding, they were human being chosen by Jesus and this placed them in a special category and a skewed distribution would probably be positive.

Since Jesus, the Master Teacher, recognized Peter's weakness and prayed for his restoration to a position of leadership, similar to a failing student who learns a "value added" lesson and overcomes his stumble to continue his academic journey. It is important to note at the empty tomb an Angel said, "Go tell His Disciples *and Peter*!" (Mark 16:7) Consequently, the distribution would be skewed with only Judas in a negative position.

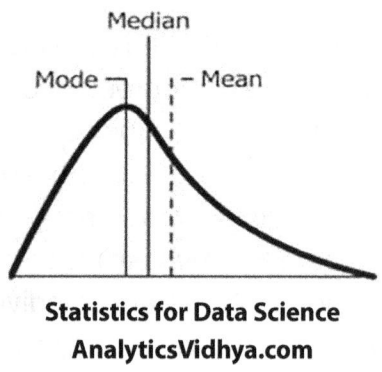

Statistics for Data Science
AnalyticsVidhya.com

This skew is similar to a final exam in a difficult subject by students who had made the long journey to the end of the course; their distribution on a final and difficult exam would have a positive skew. At least that is what a good professor would expect. After all, student failure could be a teacher problem, based on test construction which did not match class emphasis on essential course elements.

The facts are clear: the human element works against academic and spiritual development and there will be some who do not fully measure up to the expected standard in an academic environment or a faith-based entity. Although Peter stumbled badly at a crucial time in the life of Jesus, he was later restored. Obviously, the Eleven who replaced Judas with Matthias saw no reason to replace Peter but restored him to a place of leadership.

What could be learned from the above? Judas who carried the purse for the group often expressed a greediness about how resources were used. When Judas was replaced with Matthias by nomination and vote of the remaining First Disciples most likely Peter was discussed. Both nominees to replace Judas were selected because they had been with the group from the Baptism of Jesus to the Ascension. Perhaps they considered Matthias as having more knowledge of how to manage money or more likely they saw the restoration of Peter and remembered the prayer of Jesus for Peter and the final Challenge of Jesus *"As you go make disciples of all nations."* The faithful followers of Jesus were rapidly becoming aware of personal responsibility for their own lifestyle and ministry. The aggressive nature of Peter and his exchanging the material sword for the Sword of the Spirit, which was the Word of God demonstrated that Peter was an overcomer and worthy of his leadership role in advancing the message of grace and forgiveness.

Somethings are learned from books and teachers; yet many useful bits of knowledge come from others and exposure to the circumstance of real-life. Since all truth is of God, regardless of where and from whom the facts are learned, they may be stored in the long-term memory for recall later. This is why we know that all students are not equal in their learning participation or in the knowledge of subject matter at hand. This is true of those who read books or are formally taught in a classroom. Those who hear a political speech, or a faith-based homily may be classified as present, but their capacity to hear, understand and receive the contents would differ. They would be equal by definition (:=) but distributed because of other factors. There is a skewed distribution of learning due to many factors. Interest in the subject is an issue. Advance preparation to become an informed participant is another. One normally learns the unknown based on a relationship to prior knowledge stored in their knowledge base. The ability to sort out essential elements that tie a subject together and enable the recall of key facts becomes an essential part of learning. Some teachers call this grading on the curve because student attitudes or predisposition to prepare to accept and retain certain data create the spread of grades that usually appear to be skewed in a positive direction because of the selective process that places an individual in a particular field of study or advancing with a special subject cohort.

Healthy age-specific development together with positive influence from family and friends will produce an excellent student who desires to learn. Teaching such students is almost effortless because they have interest and prior knowledge and are self-starters. The development of certain personality traits and individual characteristics may produce an informal learner who requires a more relaxed and less structured process for learning. Such students may see things differently than others and develop study and work habits that may not fit the norm. In reality, these students may not receive the highest grades in the class, but they may actually be better educated with long-term benefits that predict personal achievement.

In reality, the difference between an "A "and "C" student is about one hour. The "A" student crams for an exam, unloads, makes a good score and then forgets the facts. *(How many "A" students do you know who could pass their final college exams six (6) months after graduation?)* On the other hand, a "C" student discovers why they missed an answer and learns it during review. Now they know the answer, but the system will not give them credit, because they did not know at the time of the test. This is a weakness in education and in the teaching/learning process as viewed by some faculty.

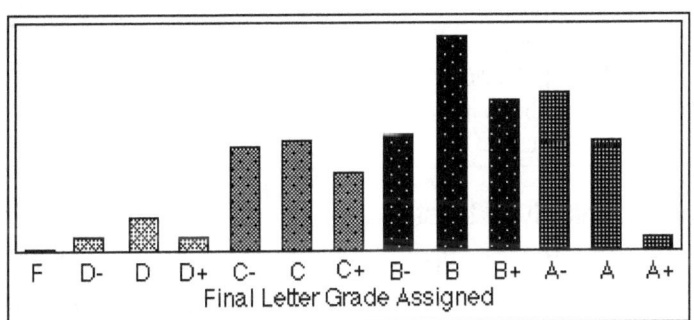

Perhaps the fact that Peter, a crude and aggressive fisherman who cut off the ear of a man who wanted to arrest Jesus, was restored to leadership after a human failure, was a lesson learned by the remaining Disciples. This teaches the value of review and application of the essential elements of a course. Later there was John Mark who disappointed Paul on a missionary journey but was later seen as a mature believer by Paul as "useful to the ministry."

> **1. Brethren, if a man should make an unintended error due to weakness, you who are regenerated, repair and adjust him with a teachable spirit; continue considering yourself, lest you also be tempted to make a false step.** 2. Practice in sharing the heavy burdens of others, and you will fulfill the principle of Christ. 3. If a man supposes himself to be something when he is really nothing, he deceives himself. **4. Let every man test himself for innocence, and then he shall rejoice in himself and not in another. 5. For every man must carry his own personal load.**

(Galatians 6:1-5 EDNT)

Obstinate Learners	Reluctant Learners	Informal Learners	Excellent Learners
Are unwilling to change and difficult to manage	Are cautious about participating, but will follow the crowd	Require a more relaxed, less structured process	Have a strong desire to learn

Conceivably, a twist of an old adage *"You can lead a horse to water, but you cannot make it drink"* could provide teachers, speakers and authors understanding of poor listeners and weak learners. One may reason from this *"One may point a student toward subject matter, but the teacher cannot learn for them or force them to learn for themselves!"* **Could this be why some learn only part of what is presented in class, shared in faith-based guidance, or written in a book or syllabus?**

To change the future of such a person, a group or an individual must demonstrate **a little compassionate concern.**

> *22. For you learned concerning your former behavior, that the old man was corrupt according to deceitful passions; 23. and the spirit of your mind must be remade; 24. and that you clothe yourself as a new man, which after God is created in righteousness and the holiness of truth. 25. Wherefore speak every man truth with his neighbor without falsehood: for we are members bound one to another. 26. Have righteous anger without sin: let not the sun go down on your anger: 27. Neither give an opportunity to the devil. 28. The thief must steal no more, but let him do honest work with his hands, that he may have something to share with the needy. 29. Let no unwholesome words come from your mouth, but only good words for enriching, that it may serve as a blessing to the hearers.*

(Ephesians 4:22-29 EDNT)

In some educational systems, the informal learner may be classified as a "C" student but will probably end up owning the business and hiring the excellent "A" students to do the work. A current list of college dropouts who became millionaires hired "topnotch graduates" to develop their ideas. A list in history of under achievers in college may validate this possibility. Some of the informal learners that ended up on top of the heap were Edison, Lincoln, Einstein, von Braun, Reagan, Kennedy, Clinton, and the list goes on and on. President George W. Bush said in this context *"I am an inspiration to all "C" students!"*

A faith-based knowledge from scripture requires a relationship with the Holy Spirit and learning from good books, good teachers and anchored by a moral lifestyle. It is the Spirit who sufficiently enlightens the sacred scripture for seekers to learn the basics. The words of the Master Teacher are worth repeating: **"The learner is not above his teacher: but everyone who is fully taught will reach his teacher's level"** (Luke 6:40). Most of us will have to wait until death and glorification to reach the level of a Master Teacher. However, the natural man cannot fully understand the Word of God; it is spiritually discovered and recognized. This supports the fact that all who claim to believe do not understand or behave the Word; therefore, some are not **fully taught**, others had weak teachers; consequently, they are not equal (=) except by definition (:=). The unconverted and those who do not *"study to be approved by God"* will be hampered by a weak teaching/learning process and are unable to be excited and directed to an adequate self-study practice both before and after exposure to new subject matter.

> *If our gospel is veiled, it is veiled to the lost: 4. whose unbelieving minds are blinded by the god of this world, lest the image of God and the glorious light of the gospel of Christ should shine unto them.* 5.*For we preach not ourselves, but Christ Jesus the Lord; and ourselves your servants for Jesus' sake.* 6. **The same God who caused light to shine out of darkness has caused His light to shine within our hearts, to give the light of understanding of the glory of God.** (2 Corinthians 4:3b-6 EDNT)

This is a weak part of the teaching/learning system in both the public domain and in faith-based operations. All mature adults have a teaching role in life, and all professional have an obligation to share their knowledge and experience with all in apprenticeship roles. When individuals do not know the basics or foundational knowledge, they cannot grasp the professional or trade knowledge when shared.

This is called the "scourge of knowledge" when an experienced professional assumes another knows and understands the basic steps. Teachers and professionals often fail to lay a sure foundation for subject matter transfer. Even experienced authors when publishing a new edition of a work will omit some basic steps assuming the reader knows them already. It is easy for a professor or a member of a social profession to assume that a newcomer knows the basics when often they do not. These assumptions are often made for new folk beginning their involvement in a faith-based operation.

> **Sidebar:** A hard lesson was learned in graduate school when several major finals were given in the same week. Since my grade point in NT Greek was high, I spent my time on weaker subjects. Tragically, the exam related to all the axioms (the established rules and principles) learned in the first semester of a four-semester course in *koine* Greek. Having moved on to translating, many of the basic language rules were not remembered. The "C" grade in my best subject taught me a remembered lesson about the basics of a subject. The same pattern is often present in faith-based guidance and sharing. This is often true in parenting when children are not taught the first principles of life or the rudimental facts of morality, ethics, and interaction with others.

An opportunity to learn becomes an obligation to study and be an informed participant in relevant subject matter. Even reading a book, a serious student should do advance thinking about the author and their qualification for writing on the subject at hand. When it comes to a structured class, the more one knows about the teacher's background and qualification to teach particular subject matter, the more confident the learner is to listen intently. The emphatic words of the Master Teacher are worth repeating **"The learner is not above his teacher: but everyone who is fully taught will reach his teacher's level"** (Luke 6:40). To learn effectively, one must have a "teachable spirit" and a commitment to study and be an informed listener with intelligent and relevant questions. Both the integrity and knowledge of the teacher are relevant to learning. Also, the learner's background, preparation, knowledge base and interest in the particular subject being taught determine the level of listening and learning.

According to Jesus, there is a level and limit to the amount and quality of what is learned because one cannot exceed the teacher's level. This does not mean the learner cannot build on what was learned from the teacher and gain additional knowledge through self-study. One thing is true: no one learns everything that is taught. Only part of what is taught is relevant to the student's capacity to build on prior data firmly established in the long-term memory in their knowledge base.

All faith-based groups in a pluralistic society are a mixing bowl of various cultures and traditions. This influences all learning and has great impact on faith-based learning, because most people try to stay in their comfort zone. The order of service, the music and the style of the pastoral homily, all are interpreted from a cultural perspective. Culture and family tradition have great influence on how various doctrines are explained and how problem passages of scripture are interpreted. Each person in the group has an obligation to balance culture and historical background with sacred scripture to effect constructive unity and understanding of faith-based living. However, the force of tradition and a cultural application of sacred teaching creates discomfort. The multitude of opinions that influence group thinking causes more division than a common expression of faith. An unclear dichotomy develops and often either instigates a group split or without unity the group simply falls apart for the lack of participation. Failure to attend or participate is considered *"negative participation"* and soon there is no one with whom to argue over polity or behavioral standards. Normally a negative position does not reach a positive conclusion.

Reading to frame my thinking for the day, it dawned on me that little was learned from history. Being a young boy when WWII started, the wartime stories and the animosity they generated, always fascinated me. It appears that America followed the thinking of Winston Churchill when he *"fused the German people and the Nazis movement into a single hated enemy."* This was done to defeat a Nazis ideology that controlled Germany (1933-1945). The same was true of the Japanese: after Pearl Harbor most people were unable to separate the Japanese war machine from the people of Japan. Often the character and behavior of some within the culture are attributed to the whole. Euclid's self-evident truism *"What is true of the whole is not true of the part"* and the converse being true also ought to be considered when one attempts to generalize the action or intent of a few to the entire group. ***What is true of a part is not true of the whole.***

> **10. But you have closely followed my teaching,** *the conduct, the purpose, the faith, the long-suffering, the love, the endurance, 11. the persecutions, the sufferings, which happened to me in Antioch, Iconium, and Lystra; which persecutions I endured: but the Lord delivered me out of them all. 12. Yes, all who will live godly lives in Christ Jesus will be persecuted. 13<u>. But wicked men and imposters will keep on going from bad to worse, misleading others and deceiving themselves.</u> 14.* **But continue to hold fast the things you have learned and been convinced of, knowing the teachers from whom you learned them; 15. and from early childhood you have known the sacred letters, the ones able to make you wise unto salvation through faith in Christ Jesus.** *16. All sacred writings are God-breathed, and serviceable for teaching, for warning, for correction, for instruction in righteousness, 17. in order that the man of God may be adequately equipped for every good work.* (2 Timothy 3:10-17 EDNT)

PART FOUR: Final Steps

Step IX: Crown Faith with Moral Excellence (Σ)
Step X: Advance a Kingdom Lifestyle ($\therefore$)
Step XI: Gain Kingdom Entrance (∞)
Step XII: Refresh your Memory of These Things ($\oplus$)

Are All Believers Equal?

The Power Of Moral Excellence

(Σ)
Step IX

Crown Faith with Moral Excellence

(Σ) Symbol for sum total

***5. And you too have to contribute every effort on your own part, crowning your FAITH with** [1] (benevolent love) and [2] (brotherly kindness) and [3] (godly worship) and [4] (enduring steadfastness and [5] (self-control (6) Knowledge from books and teachers, and **(7) moral excellence)**.*

(2 Peter 1:5-7 EDNT)

And He gave some to be messengers, and some preachers, and some missionaries, and some teaching pastors; for the ultimate purpose of equipping the saints for the work of serving, for the building up of the body of Christ: until we all attain the same faith, and the experiential knowledge of the Son of God, unto mature manhood, unto the full measure of development in Christ: that we no longer behave as young children, driven before the wind of each new teaching, by the trickery and sneakiness of men, whereby they ambush with deceitful schemes; but arriving at truth in love, you may grow up into Him in all things,

(Ephesians 4:11-15 EDNT)

Basic salvation is grounded in the Word of God but is also a process of growth in experiential knowledge of God working in all aspects of life. This process begins with faith, but each convert must add to their faith various steps and grow until the level of moral excellence is reached. Conversion has to do with past sins and the past life. One does not immediately jump from convert to a full-grown mature believer. They must learn to trust, learn to forgive, learn to share, learn to worship and the list goes on until a new lifestyle is developed and is observable. So, then, let us leave elementary teaching about Christ behind and press on to full growth; no need to lay the foundations all over again (Hebrews 6:1). Paul further instructed early believers not to return to lifeless observances:

> *1. So, then, let us leave elementary teaching about Christ behind us and pass on to full growth; no need to lay the foundations all over again, the change of heart which turns away from lifeless observations, the faith which turns towards God, 2. of the instructions about different kinds of baptisms, about the laying on of hands, and of resurrection from the dead, and upon that sentence that lasts all of eternity. 3. God willing, this is our plan. 4. It is impossible to renew to repentance those who were enlightened, those who tasted the free gift from heaven, and those who were partakers of the Holy Spirit. 5. And have tasted the goodness of the word of God, and felt the powers of the world to come, 6. if they fall away, they cannot attain repentance, seeing they crucify the Son of God a second time, and are exposing Him to public contempt. 7. For when the earth drinks in the rain that comes often upon it and when it brings forth herbage useful to those who work the ground. It receives a share of the blessing from God; 8. but if the earth produces thorns and thistles it has lost its value; a curse hangs over it, and it will feed the bonfire. 9. But we are confident of better things of you, beloved, things that go together with salvation. 10. God is not an unjust God that he should forget all you have*

done as a labor of love that you displayed in that you have been and still are active in the service of God's dedicated people. 11. But our great longing is to see you all showing the same eagerness right up to the end: 2. so that you do not become lethargic but imitate those who through faith and patience inherit the promises. (Hebrews 6:1-12 EDNT)

Why do believers have such difficulty with the moral issues of today? Partly because believers remain human beings living in a sinful and demoralized world. Tragically, world conditions are getting worse and worse as the End Time is approached. The hard lessons that come from parents are never fully understood by children; neither are the trials and tribulations that befall believers in their daily lives accepted as part of God's guidance back to the true course that leads to Heaven. Because God loves His children, He disciplines them in ways and by means that are not enjoyable. Usually, the problem is as simple as "We get the cart before the horse." In other words, we often see things backward to the way God presents them to us.

Notice the Greek emphasis is in reverse from the English order. Human willpower or New Year's resolutions will not correct the errors of life or construct a more positive and productive lifestyle. Moral excellence is gained over time through a specific process which must be enabled by the Holy Spirit. Mankind in the human state cannot develop moral excellence without faith and faith comes by hearing, understanding, and accepting of truth. There must be attentiveness, listening, analysis, and action that comes from a reverence and respect for sacred truth. Respect, meaning *"to look at and pay attention to"* the person and/or subject at hand. **Moral**

excellence is the last rung in the ladder of grace and knowledge that enables one to develop a missional lifestyle. The other steps must be reached before moral excellence is attained and a missional lifestyle is observable by others or productive for constructive social change.

Peter reminded believers that the common salvation with all the benefits and blessings required repentance and the acceptance of both experiential and personal knowledge of how God works in individual lives. It is not book knowledge or shared subject matter gained from teachers; it is knowledge from personal experience and first-hand awareness that God is acting in us, around us and through us! Answered prayers and seeing God work in the lives of others adds to such full or experiential knowledge which the Greeks called *"epigon'ei"* as opposed to *"gno'sei"* which was knowledge from books and teachers. When Peter speaks of *the common privilege of faith* shared with other believers along with justification, grace, and peace which comes by the experiential knowledge of God, he is speaking of a process of growth and development...a time of becoming!

To crown faith with **moral excellence**, we must begin with (1) **benevolent love**. Then, we add (2) **brotherly kindness**, followed by (3) **godly worship**. Now, comes the hard part, we must add to our faith (4) **enduring steadfastness** and (5) **self-control**. Next is the addition of (6) **knowledge from books and teachers**, and finally believers arrive at (7) **moral excellence**. This is the key that opens the door to a moral lifestyle and an **abundant entrance to the Kingdom**. Salvation is not by works, but by the work of the Holy Spirit who convicts of sin, righteousness and judgment. Redemption for the human soul comes when one agrees with the verdict of scripture and accepts the grace of Divine forgiveness.

Salvation is God's free gift but living the life of a believer requires continuous positive action by taking the steps of faith which facilitate growth and knowledge of a missional lifestyle which is kingdom living. Notwithstanding, that salvation is the free gift of grace, believers must contribute their part to their spiritual growth. The suffix *ion* adds *action* or *condition* to the words: *salvation* and *redemption* both require some action on the part of the receiver. *The acceptance of God's saving grace is a start of a change of action, attitude and lifestyle.* No one can do this for someone else; these steps must be taken by the individual this spiritual process. There can be no reluctant believers; all must be active learners of grace and spiritual knowledge.

Believers must start at the bottom of the staircase or ladder with **benevolent love** and climb upward toward **moral excellence**. With perseverance, truth, and love woven into the fabric of faith as the garland of an overcomer. According to Jacob, the place of the staircase or ladder was the House of God and the gateway to heaven (Genesis 28:17). This shows the value of gathering together with other believers to worship, celebrate and share a common faith in fellowship.

According to 2 Peter 1, the seven steps in the staircase or ladder that the believer must climb to **crown their faith with moral excellence** is in the order of Greek emphasis (7-1).

7. **Moral Excellence.**

6. **Knowledge from Books and Teachers**

5. **Self-control**

4. **Enduring Steadfastness**

3. **Godly Worship**

2. **Brotherly Kindness**

1. **Benevolent Love**

The spiritual life is not one of trial and error; one does not partially accept God's mercy and grace. There is no provision for part-time service to God, either one serves God or travels the way that leads to perdition. A firm "choice" must be made. It does no good to **"try on God's armor for size"** and then discard the protection before the battle for life is won. Putting on Sunday clothes and attending service is not sufficient, one must **"wear the whole armor of God"** into their daily battles as a good soldier of the Cross. Paul told young Timothy, *"Join the ranks of those who share hardships as a soldier of Jesus Christ. 4. No active warrior entangles himself with ordinary affairs; as a result, he may please the one who enlisted him as a soldier."* (2 Timothy 2:5-7 EDNT)

> *And you too have to contribute every effort on your own part, crowning your faith with (7)* **Moral Excellence**, *and to moral excellence (6)* **Knowledge from Books and Teachers**: *and to your knowledge (5)* **Self-Control**; *and to self-control (4)* **Enduring Steadfastness**, *and to enduring steadfastness (3)* **Godly Worship**; *and to godly worship (2)* **Brotherly Kindness** *and to brotherly kindness (1)* **Benevolent Love**.

(2 Peter 1.5-7 EDNT)

Notice that these stages are in reverse order from the English listing in KJB. This is because the Greek emphasis begins with benevolent love *(agape)* or one-way love, the same way God loves us, and ends with moral excellence which is beyond the reach of sinful man without redemption. It is obvious from a casual observance of the real world that immorality and wickedness is rampant and *"being in the world and not being part of the world"* is a divinely enabled process but there must be diligent effort on the part of each believer. The righteous life does not drop out of heaven as the rain nor can it be developed overnight. God is constantly working on each of His children to bring them to moral excellence and the process begins with (1) benevolent love, which comes from God, *"because the love which God has for us is poured out in abundance in our hearts by the Holy Spirit which is God's gift."* (Romans 5:5)

Once a believer establishes a predisposition for (1) **benevolent love** expressed toward others the way God loves, they must develop (2) **brotherly kindness**. This is the two-way love of friendship from which we get the word Philadelphia *"the city of brotherly love."* Believers are ready for (3) **Godly worship**; that is, to show respect and worth-ship of God in their lives. What is normally called worship is in reality the *"worth-ship of God."* (Review Step VI). How does one value God in all aspects of their lives not only on Sunday morning. Worship is an attitude and should be the continuous mindset of all believers in all their waking thoughts. In reality, worship is the cargo ship that carries the worth and value of God as the believer travels the dangerous seas of life.

An honest acknowledging of the worth and value of God in all aspects of a believer's life is the foundation stone for achieving (4) **enduring steadfastness.** Not occasional steadiness, but a persisting and sustaining faithful devotion to God and a total commitment to the cause of righteousness. The sustained devotion to God that brings steadfastness logically leads to (5) **self-control**. This comes with maturity which establishes self-discipline, restraint, strength of mind and will. Self-control is a first cousin of maturity. Self-control becomes maturity when one mellows from experience which causes reliability, responsibility, and wisdom beyond their years. According to Dear Abby's mother. **"This is maturity: To be able to stick with a job until it's finished; to do one's duty without being supervised; to be able to carry money without spending it; and to be able to bear an injustice without wanting to get even."**

Next comes (6) **knowledge from books and teachers**. This is *"taught and learned"* knowledge from material sources contrasted with *"true experiential knowledge"* which comes from personal experience. (2 Peter 1:1) deals with the "common privilege of faith and grace and peace that is multiplied through full knowledge of God and Jesus Christ our Lord." Full knowledge is gained experientially based on personal interaction or observed behavior and/or a shared mystical experience with God and leads to step (7) **moral excellence**.

Once a believer has personally experienced God's intervention in their life and has accepted guidance through the stages of growth in grace and learned from others the basic knowledge of a moral lifestyle, they are ready to crown their faith with (7) **moral excellence**. Experiencing the full knowledge of God through answered prayer, personal spiritual blessings, and observing God working in the lives of others in a place of worship, they are ready to demonstrate their faith through a missional lifestyle and validate the worth-ship of God in their life. This is the crowning achievement of a Believer's lifestyle produced by moral excellence. Having experienced Gods intervention in their life, they are ready to entrust what they have learned to others who will be competent to share with others the message of grace and eternal life. Moral excellence through God's intervention is the correct Crown for their faith and provides the integrity to share with others both the easy and the deep things of God.

> Such gifts, when they are yours in full measure, will cause you to be neither unproductive nor unprofitable in the full knowledge of our Lord Jesus Christ. He who lacks these are no better than a short-sighted man feeling his way about; And has forgotten that his old sins have been purged. So, believers, be the more eager to confirm your calling and your choice: for if you do practice these virtues, you will make no false steps: and you shall be richly supplied the entrance into the kingdom of our Lord and Savior Jesus Christ. (2 Peter 1:8-11 EDNT)

Another case of reverse order is found in Luke (10:25-27). A young lawyer asked Jesus what he had to do to inherit eternal life. Jesus answered with a question *What does the book say?* The answer from the book which the lawyer knew by heart and Jesus commended his knowledge and said, *"This do, and you shall live!"* But did the man understand the proper sequence? Converts do not begin their spiritual life fully grown, but as **newly born babes** and must grow and develop into mature believers. They must learn to believe, learn to trust, learn to obey, learn to sacrifice, and learn to be an overcomer. "Faith born of God can conquer the world." (1 John 5:4).

The book order in Luke was love God with your whole (1) **heart**, (2) **soul**, (3) **strength**, (4)**mind**, (5) and your **neighbor** as (6) **yourself** but this is not the Greek order. Just as one cannot first reach moral excellence without following the proper sequence; one cannot learn to continually love God with the **whole heart** without the developmental process. (1) You must **first love yourself**; (2) Then **you love your neighbor** to the same degree; (3). Now, you must **love God with your mind**; (4) Then your **whole soul** (mind-will-emotions); (5) your **strength**, and (6) Finally, you **learn to love God with your whole heart**, which is the control center for your life which brings one to (7) **a moral lifestyle**. (Luke 10:25-37 EDNT)

> .... *until we all attain the same faith, and the experiential knowledge of the Son of God, unto mature manhood, unto the full measure of development in Christ: that we no longer behave as young children, driven before the wind of each new teaching, by the trickery and sneakiness of men, whereby they ambush with deceitful schemes; but arriving at truth in love, you may grow up into Him in all things.*
>
> (Ephesians 4:13-15b EDNT)

(∴)
Step X

Advance a Kingdom Lifestyle

(∴) *Symbol for therefore*

Such gifts when they are yours in full measure, ***will cause you to be neither unproductive nor unprofitable in the full knowledge of our Lord Jesus Christ.*** *9. He who lacks these is no better than a short-sighted man feeling his way about; and has forgotten that his old sins have been purged. 10.* ***So, believers, be the more eager to confirm your calling and your choice: for if you do practice these virtues, you will make no false steps:*** (2 Peter 1:8-10 EDNT)

The writer of Hebrews shared with those who had accepted the Word of Faith, believed and were learning to behave, words about those who remained tied to the old ways. He wrote of not being delinquent and coming up short, **because the Word was received but had not been woven into the fabric of their faith when it was spoken.** (Hebrews 4:8) This meant others who heard the Word and refused to believe and held on to old traditions that had been superseded by the death of Jesus, the fabric of their faith was not sufficient to cover their old lifestyle and allow a new way of life to develop.

Let us be on guard, while the promise of entering His rest still holds; that ***none of you may be found to be delinquent and come up short.*** *2. The good news was proclaimed to us, as well as to them: but* ***the word was not heard and therefore did***

not profit them, because it was not woven into the fabric of their faith when it was spoken. (Hebrews 4:1-8 EDNT)

The Kingdom lifestyle is missional, moral and ethical. Since God is Holy, His followers must be committed to an observable difference in lifestyle that demonstrates a new way of life. The change must be noticed or perceived. What is a missional lifestyle --adopting the thinking, behaviors, and practices of a called and committed follower of the New Testament plan: *"as you personally go into all the world make disciples, etc."?* The objective is not only personal entry into the Kingdom, but participation in a global outreach of grace and mercy. Hopefully, this outreach would include family, friends and even enemies that were converted and the Word woven into the fabric of their lifestyle. Then there would be abundant entry into the Kingdom, having taken all the steps that increased both the readiness to witness to God's saving mercy and work with God in reaching the world with the message of grace.

Missional living is working together with God in advancing the message of redemptive grace in the place God plants you. Wisdom brings authenticity and genuineness to the daily lives of those desiring a missional lifestyle. **Missional living is the adoption of the attitude, thinking, behaviors, and practices of a missionary in order to engage others in the process of advancing the gospel message.** In Psalm 8: 32-35 wisdom speaks further to the faithful who attend with interest to instruction and are blessed by keeping to the proper pathway. There is a warning to those who disregard the lessons learned. Lifestyle opens the door of wisdom and soul-winning (Proverbs 11:30), the fountain of blessings and the Gates of Heaven.

Those who listen and are watchful daily at the open door of wisdom will find life and favor from the Lord and will enjoy the spiritual reality of God's Presence. This is the day for fellowship among the band of believers and time for full dedication to a global outreach. After leaving Thessalonica, Paul journeyed to Corinth. Later he wrote the believes at Corinth about working together with God to advance the gospel. He basically told them *"God is working; you must get together."*

8. Now he who did the planting and the one doing the watering are part of the same process: and every man will receive a reward according to his work. 9. ***For God is working and the laborers are together:*** *you are God's farm; you are God's field to be worked and God's building to be constructed. 10. According to the favor of God given to me, as a wise master builder, I* ***have laid a foundation, and another will build on it. But let every worker take heed how he builds on the foundation. 11. There is no other foundation for the building but the one laid on Jesus Christ.*** (I Corinthians 3:8-11 EDNT)

1. As we work together with God, we appeal to you not to accept the grace of God and let it go to waste. *2. (God said, I have heard your prayers at a convenient time, and in the day of salvation I have brought you relief in a difficult situation:* ***observe, now is the time for coming together; now is the day of deliverance.****) (*

2 Corinthians 6:1-2 EDNT)

A kind of "moral algebra" is a construct that can enhance an ethical lifestyle of a family, because algebra is a process where one learns to solve problems by observing the parts. The word "algebra" means the *"reunion of broken parts"* and is the study of rules of operations, relations, and the constructions and concepts informed by these rules. There is an algebraic concept in the moral practice of a faith-based family. Basic algebra lessons emphasize the practice or action of understanding the question and intelligently and simply arriving at an answer. Just as students dislike the study of elementary algebra, troubled families are normally reluctant to study the rules and regulations of faith-based money matters and behavior that could solve their problems and open the windows of heaven. The rules and guidance in Scripture relative to family life and moral excellence are similar to a moral quandary: materialism, carnality, and the human condition pushes one direction and the spiritual dimension of faith-based living points to a better way. When the human equation is out of balance, all aspects of human existence suffer from unsteadiness.

There is a distinct lifestyle for believers based on a missional reality that supports a life of faith. Essentially, a missional reality coalesces around a personalized grasp of scripture that offers a theological shift, a sociological direction, and a distinct lifestyle for believers. The missional mindset is placed in the context of viewing the Cross through the Empty Tomb, seeing culture as a vehicle of sharing, the House of Prayer as a force to work with not of a field in which to work, because the community and the world is the mission field ready for harvest. Sadly, the workers are few! A prayer that survived from the earliest days of Christianity is worth remembering:

> Lord, be with us this day,
> Within us to purify us;
> Above us to draw us up;
> Beneath us to sustain us;
> Before us to lead us;
> Behind us to restrain us;
> Around us to protect us
>
> AMEN!

Soon after five (5) missionaries were killed in Ecuador (1967) that transformed a tribe and shaped evangelical missions for half a century, my visit to the site to write an article created a significant change in my attitude. My letter to Nate Saint's sister, who continued to work as a missionary to the tribe, asking "how" she could function as a Missionary in the place and with the people who killed her brother. My mistake was to capitalize "Missionary" and her return letter included a reprimand **"Do not ever capitalize the word missionary again. It is not the special task of a few; it is the work of every believer on every step of their journey."**

What does a missionary know and how do they feel about their task? They know beyond a doubt they have been called to serve outside their comfort zone. They understand they must leave family and friends and travel into a strange land. They are aware of the new language and culture they will face. Appointed missionaries know they must live a life worthy of financial and prayerful support from an extended constituency. They know they have limited resources and that through deputation they must raise funds to replace what is spent, or they cannot continue their work. This provides a missionary family a totally different perspective on money matters than a state side family involved in ministry.

A family involved in missions cultivates a positive mindset that God is in charge of their lives and ministry. Missionaries must teach their children to live on a limited budget and that every cent saved enhances their chance of winning a soul for Christ. In fact, the missionary and their lifestyle are often lonely and full of daily difficulties. Without safe living quarters, clean sheets on the bed, good food and water, and with only local native people protecting them against hostile forces, missionary families develop an uncertain way of life. Can we say, "God bless the missionaries!" Then in the next breath say, *"Lord help me to walk the right pathway and demonstrate a missional life-style to others and be supportive of those called to serve overseas."*

> **Sidebar:** A missionary's work is hard and often harsh, and it is even more difficult for their family. The convenience of a well-furnished home, good schools for the children, and a well-stocked food market, good transportation, and good roads are nowhere to be found where missionaries live and work. Recently, my oldest son, Barton, had an opportunity to visit Guatemala for a few days. He thought he was going to sing, pray and listen to missionaries speak.
>
> Traveling on a mountain road to an isolated village, to his surprise, he was handed a hammer and told **"We are building houses this week."** Another surprise, they were building on the side of a volcano (which erupted a few days later and killed many). In the past he had complained about my mile-long road up Lone Mountain to my retirement compound, but after this new experience he sent a message, "Tell Dad

he has a wonderful road up his mountain." It would be great if every overweight, over paid member of a faith-based congregation could spend a week on a primitive mission field living the life of a missionary worker. Most likely they would return with a changed attitude, as one of my previous members prayed after returning, **"Lord, I will freely support those missionaries, but please don't ask me to live and work there."**

10. For the one wishing to love life and see prosperous days, let him avoid an evil tongue and cunning words. 11. Habitually avoid evil and do good; let him seek and follow peace. 12. Because the eyes of the Lord watch over the righteous, and His ears listen to their prayers: *but the Lord looks directly into the eyes of wrongdoers.* (1 Peter 3:10-12 EDNT)

It is relatively easy to take a "mission trip" with a group of excited travelers, but living and working in the same austere and bleak environment over time is a different story. I remember a 300-pound pastor riding a small burro on a mountain trail in Central America. His feet almost touched the ground, and he began to feel for the small animal laboring to carry him and his luggage up the steep trail. He got off the burro but left his heavy suitcase tied on the animal. After a short walk up the rugged trail, he said to a friend **"I think I'll just get back on this donkey and help him carry my suitcase."** It was a sight to see: this super-sized preacher straddling a small donkey holding his suitcase to help the little burro with the heavy load. It is doubtful that a picture of this monstrosity would raise any funds for missions. I was just thinking, what if God had enabled that donkey to speak to the preacher as happened when *"A dumb donkey spoke with a human voice and ordered the* *prophet not to speak falsely."* (2 Peter 2:16 EDNT) Had that happened the preacher would have had a whopper of a missionary message when he returned to the comfort of his pulpit: **"While on this mission trip, God spoke to me through a donkey!...etc."**

> *22. You must be honest with yourselves and live by the word not merely hear it. 23. But those who listen to the word, and do not behave it, are similar to a man seeing his own face in a mirror; 24. he observes his flaws, and immediately forgets the man he saw. 25. But whosoever bows down to observe the complete prescriptive usage and the unrestrained opportunity to continue in the word and not become a forgetful hearer, but <u>one who behaves the prescribed deeds, this man shall by the blood be set apart for consecrated action.</u> 26.* ***If any man among you seems to be devout, and restrains not his unnatural language, he deceives his own heart and his service to God is ineffective.*** (James 1:22-26 EDNT)

James wrote about a pure and unadulterated lifestyle that acted with a missionary perspective. Sanctified eyes that could see each child without the influence of a believing father as being in need of surrogate nurturing. They could observe women alone who lack a husband in their hardship or misfortune and were in need of support and care of an extended family or a faithful friend. In addition to doing good for orphans and women alone, those with a missional lifestyle could see spiritual opportunity in each observable circumstance. True evangelism is meeting the needs of hurting individuals and families at the earliest point in time at the farthest distance from a place of worship as one is guided by the Holy Spirit.

Thinking how a missionary normally lives and works could cause one to see opportunities where others simply see troubles and trials. In the presence of family disfunction and personal difficulties there are opportunities to lead others into the Place of Prayer for soul cure and spiritual fellowship. This can truly be "snatching souls" from the hands of Satan and bringing them into the Arms of Jesus.

Do we need another *"come to Jesus moment"* for a second touch? All believers need clear vision to see the opportunities placed before them daily. Most churchgoers could use a "second touch" of Jesus to correct their short-sighted vision of both humanitarian and spiritual needs.

We need to focus on the present needs of those within our sphere of influence. The familiar story of a wounded man left by robbers on the roadside who religious leaders passed by without offering assistance, while a traveling Samaritan both saw and acted on behalf of the urgent personal need of a stranger. (Luke 10:25-37)

There would most likely be more true celebration at places of worship if more believers behaved as did the Samaritan rather than conducting themselves as others who both saw the problem and neglected to act at the point of need. Perhaps a second touch by Jesus would produce a moment of spiritual reflection? Some start the journey and others take a few steps but need "a second touch" by Jesus to see clearly their opportunities and obligation. Mark records a blind man being brought to Jesus.

23. And Jesus took the blind man by the hand, and led him outside the village; and dampened his eyes *with saliva and laid on His hands, and asked him can you see anything? 24. And the man looked up, and said, <u>I see men as trees, walking.</u>* **25. After that Jesus laid His hands on his eyes again, and his sight came into focus, and he saw every man clearly.** (Mark 8:23-25 EDNT)

The Practice of intentionality in lifestyle is a deliberate and planned process. Yet there are factors that hinder a progressive walk with God. The American and English driving practice is a good example of a difference in understanding the left and right. Americans drive on the right side; the English drive on the "correct" side. Another example would be the word "momentarily." For an American the word means "in a moment" for the English the word means "for a moment." Should the Airline Pilot say, "We will be taken off momentarily" the meaning may be understood differently. The same seems to be true for some of the political language in America. Some are on the "left" and others on the "right."

What does "left and right' teach us? – Being a clergy/educator, a normal approach would be to go to the original language of the New Testament (Greek) to understand left and right. Matthew 6:3 explained in alms giving "don't let the left hand know what the right had is doing." And Luke 23:33 the thieves were placed on either side of Jesus; "one on the right hand, and the other on the left." This supports that charity gifts should be done in secret and when it comes to politics there are criminals on both sides and the "good" is in the middle. For left the Greek used *aristeros* meaning "second best" and for the right *dexios* meaning "dexterity or right side." The left was the side less used and the right had dexterity that cannot be acquired without much practice and experience. So what does the "left and right" teach us? Probably nothing; the good is somewhere in the middle with bad forces on both sides. This is why a moral renaissance is needed. And why a missional lifestyle is required for believers.

Understanding "we" and "me" – Consider your left hand and your right hand. The left hand is "we" and the right hand is "me." The ring finger is on the left hand in Western culture; therefore, the wedding band is normally worn on the left hand because it signifies a relationship with another. Whether it is politics – left and right, or social or spiritual change, the difference can be explained in terms of "we – the group" or "me – the individual." Social change and positive integration will take place at the individual level first; then it can proceed to groups and the community.

Tragically, many faith-based splits were simply efforts to restore unity or fellowship around a common teaching. The development of a new brand-named group is evidence partitioning is still hindering the Gospel message from reaching the world. Some present-day groups did not begin with an intent of starting a new denomination but were seeking restoration of New Testament behavior. Church history recorded various efforts to turn back the tide of liberal interpretation and the drastic change in moral behavior and lifestyle. Such efforts as the Gospel Union, the Restoration Movement, the Reform Theology effort, an association of local congregations, ministerial alliances, collective groups of congregations who share similar beliefs, cultural and regional groups who impose their culture and tradition on faith-based practices, and the variety goes on and on.

This process has muddled the message of grace until religion has little influence on society. We are back to individual faith and family-based groups. The big question: *"When Jesus returns will He find faith on the earth?"*(Luke 18:8) The text here implies "Will men have faith and be praying when The Son of Man returns?" This matter is still in question. With the progressive debauchery and wickedness getting worse and worse with continual deception, with declining attendance in Bible study, and listeners for the Sunday homily ... drastic correction is needed in the worship and lifestyle of believers.

Faith-based principles can make a difference in a moral lifestyle. Starting with the Commandments from the Hebrew Torah and appearing in various forms in the scared writings of most religions, common sense lessons are taught by faith-based scholars. One such lesson is recognized as a Golden Rule and moral leaders are stewards of both the instructions and the resources provided by Providence and preserved in sacred writings. The guidance of this Rule is so basic it becomes an attitude – a predisposition to behave with fairmindedness and impartially.

Attitude is a predisposition that is assessed by action, while behavior is a measurable activity. Latin words are the secret to understanding those who contest the status quo and others who protest for constructive social change. The struggle of history's Reformers was to reform and move people to follow moral principles rather than the whims of men. They were called "protestants" because they were more for constructive change in the lives of believers than they were against the established institutions. These were communicated in sacred writing in various forms: seeking change to improve not destroy.

Luther was a professor of moral theology at the University of Wittenberg who raised an academic disputation about the efficacy of indulgences and posted his concerns for the morality of the people who believed they could purchase the privilege of indulging in immorality. His objective was moral purity of the people, but the church leadership refused to accept Luther's scriptural clarification: *"The just shall live by faith"* and continued their lenient attitude toward pleasure-seeking and self-indulgent behavior of parishioners. Luther's work was both a plus and a minus for universal faith-based worship.

> **Sidebar:** Major World Religions have behavioral rules and principles to guide relationships:
>
> **Buddhism** –"Hurt not others in ways that you yourself would find hurtful." *(Udana-Vaarga 5,1)*
>
> **Christianity** –"As you would that men should do unto you, do you also to them likewise." *(Luke 6:31)*
>
> **Hinduism** – "This is the sum of duty; do naught unto others what you would not have them do unto you." *(Mahabharata 5, 1517)*
>
> **Islam** – The ethics of reciprocity, is an Islamic moral principle which calls upon people to treat others the way they would like to be treated. Although it is not mentioned in the Quran, the principle was stated many times by the Prophet Muhammed.
>
> **Judaism** – "What is hateful to you, do not do to your fellowman. This is the entire Law; all the rest is commentary." *(Talmud, Shabbat 3id)*
>
> **Taoism** – "Regard your neighbor's gain as your gain, and your neighbor's loss as your own loss." *(Tai Shang Kan Yin P'ien*

Good families make the difference in all aspects of a civil society. It was good parents and good families that became the foundation stones for a moral society and the building blocks to structure faith-based living beginning in Jerusalem. The first men chosen as Lay Leaders in the Jerusalem Assembly (Acts 6:1-15) were honest family men with well-disciplined children and a good report in the community. Home-based schooling is not a new addition to education, mothers and fathers have operated home-based education for centuries. The obligation of faith-based behavior begins with the sharing of Divine mercy and grace with family and friends with a priority example for the disadvantaged poor and dysfunctional families.

Paul expressed both attitude and action to establish and express such affection for his converts, functioning with the love of a mother and the discipline and guidance of a caring father: (1 Thessalonians 2:7-12 EDNT)

Behavior of a Mother

7. But we were tender among you, **even as a nursing mother warmly takes pleasure in her children:** *8. so affectionately longing for you, we were willing to share with you, not only the gospel of God, but also well-pleased to share our lives, because you were valued by us.*

Behavior of a Father

9. You remember our long and hard labor night and day, because we would not burden you for expenses, but freely preached the gospel of God unto you. 10. You are witnesses and so is God, how upright, honest and blameless was our conduct among you that believe: **11. as you know how we encouraged, comforted, and charged every one**

of you, as a father treats his children, *12. that you would lead a life worthy of God, who has called you unto the glory of His kingdom.*

Sidebar: Grace Irene Curton was born on the wintery morning of February 11, 1905, the third child of Ida Aldona Dobbs and Robert Tate Curton. She became a teacher in Rhea County Tennessee, and a chance meeting made her the bride of Herbert Barton Green. Grace also taught Sunday school and sang in the choir most of her life and served as Dorm Mother for Nurses-in-Residence in Ohio and Pennsylvania and retired (age 67) as Dean of Women at Lee College.

In her later life, she traveled extensively in the US, to the Caribbean, and four trips to Oxford, UK. On one occasion she brought back a suitcase full of books, when the Custom Agent lifted the heavy case and asked, "What's in here?" Mother answered, "Books, I read a lot!" She read to keep her mind active and it worked.

In the early morning hours of May 4, 1996, she finally realized the truth of her long-held philosophy: ***"Never be afraid to trust your unknown future to the all-knowing God,"*** and she once wrote, ***"If God sends us on rocky ground, He provides strong shoes."*** As the mother of three and a widow of 59 years approached the threshold of reunion with Barton, she was alert and at peace. Her last words are not only a tribute to her life, they are a testimony of her faith; teaching those who remain, the ultimate Lesson of Grace; *"I am not afraid. I am ready to go."* - Excerpt from her Eulogy by a grandson, Barton L. Green

The love of motherhood graced our fatherless family for 59 years. My father passed in 1937 and widow Grace, with the love of a mother and the total care of a father, assumed the dual role for three fatherless children. The pain of loss remains, but the memories are good.

There is no replacement for a mother's love and concern (unless God's love is considered).

The family is a training ground for a civil society, leadership development for faith-based operations and the integration of moral excellence into the lifestyle of believers. Children are to be taught to respect others and share with the less fortunate. Perhaps we should all develop such a respectful concern for the disadvantaged. The young need both a mother and a father. When this is lacking, the surviving parent must do double duty. Each human being has the capacity to **share benevolent love and demonstrate a moral excellent lifestyle.** This not only requires **"the love of God shed abroad in a believing heart by the Holy Spirit"** but also the exhibiting of both maternal and paternal qualities toward the young. This is needed to guide the young into faith-based behavior that ultimately leads to **an abundant entrance to the Kingdom.**

We sing "When we all get to Heaven",

but in the "we all" there is a "me."

Perhaps we should also sing,

"When the Saints go Marching in

I will be in that number."

If you practice the believer's lifestyle,

You will make no false steps:

And you shall be richly supplied the

Entrance into the Kingdom!

Are All Believers Equal?

The Power Of Moral Excellence

∞

Step XI

Gain Kingdom Entrance

(∞) Infinity or unlimited number

8. Such gifts, when they are yours in full measure, will cause you to be neither unproductive nor unprofitable in the full knowledge of our Lord Jesus Christ. *9. He who lacks these is no better than a short-sighted man feeling his way about; and has forgotten that his old sins have been purged. 10. So, believers, be the more eager to confirm your calling and your choice:* ***for if you do practice these virtues, you will make no false steps: 11. and you shall be richly supplied the entrance into the kingdom of our Lord and Savior Jesus Christ.***

(2 Peter 1:8-11 EDNT)

The kingdom journey is physically and emotionally hostile to the human nature. Satan takes advantage of each hardship and difficulty to discourage and misdirect. Although it is a spiritual journey on a path planned by God; it is an obstacle course which must be navigated by faith and negotiated with courage to surmount the obstructions. The race must be finished to be classified as an overcomer. Satan remains the Prince of the earth and controls a majority of humanity. It is a battle between right and wrong: a battle that must be won. It is an earthly journey on a pathway designed by God which is straight and narrow and passes through hostile territory. It is a battle that is not over until it is over. Death is the ultimate victory. All must learn to be a good soldier and wear the complete Armor of God - not just try it on for size. Compared to the sacrifice of Jesus, suffering and sorrows of early disciples and the daily hardships of present-day believers, the trials and tribulations of the present are not worthy to be compared to the Crucifixion that brought mankind eternal life or the eternal glory at the end of the journey (Romans 8:18). Yes, there are times of blessings, peace and rest when the mind is focused on Jesus. The book of Hebrews lists those who gave their lives but completed the journey. The writer of Hebrews identified those who through faith became overcomers and ended with spiritual victory even though they lost the physical battle. Death became the ultimate victory. **"For us, God had something better in store. We were needed to make the history of their lives complete."** (Hebrews 11:1-40 EDNT)

The struggle of John Bunyan, and his (1678) classic faith-based allegory "THE PILGRIM'S PROGRESS," was about **a pilgrim traveling from this world, to the world which is to come.** Bunyan felt called to ministry but would not accept the endorsement of the Government for Licenses. He was placed in prison and made lace to support his family and exercised his ministry through writing. His work is regarded as one of the most significant faith-based writing in English literature (new converts should read this book more than once).

> I loaned 25 books to a man on Death Row that included a modern version of *Pilgrims Progress*. After his execution, the prison Chaplain returned all the books but *Pilgrims Progress*. He wanted all those on Death Row to read it. Finally, it was

returned with obvious wear. Bunyan's classic remains a force for believers.

The most important episode in the book, from this author's perspective, was when Christian was about halfway up the Hill of Difficulty and met two fellows running down. They were named Timorous and Mistrust and they yelled, **"Don't go up there; there are lions up there!"** Christian responded, **"To go back is nothing but death, I will go forward!"** There were lions at the top, but as Christian negotiated the straight and narrow pathway, he could see the lions were chained and could not reach him on the path. Also, he saw they were old, and their teeth were rotting. He had climbed the Hill of Difficulty, overcame all obstacles, faced the lions, and knew he was safe on the straight and narrow pathway to the Eternal City.

Adversity can be the foretaste of faith-based achievement. No mountain is too high, no valley too low and no cost too great but that diligence and perseverance can bring abundant entrance into God's Heaven. God made both the mountains and the valleys, then man was made from the earth and told to *"subdue, conquer, overpower, and overcome"* life's difficulties. Consequently, mankind throughout the generations learned to surmount hardships and grow a garden, build a home, raise a family, start a business, develop a community, establish faith-based operations and lead nations. The accumulated wisdom of the ages is preserved for those who wish to achieve a constructive purpose. Bunyan did not have a pulpit, but PILGRIM'S PROGRESS was translated into over 200 languages and has never been out of print.

It was my privilege to study "Military Science and Tactics" for three years as a military Chaplain during the Vietnam era. I made many applications to the Christian life. Following Napoleon's victory strategy was a good military tactic: when both sides have fought to a stalemate, the one with courage to attack after that point will be victorious. There is a clear lesson in this strategy, when trouble comes one must take positive action to move past the difficulty. Once General MacArthur was asked about a retreat he ordered and he corrected the record, *"It was a strategic withdrawal to fight another day."* Loss can create an opportunity for gain. This common lesson from the past is most telling, *"Necessity is the mother of invention."* Difficulties may become stumbling stones and produce failure or become stepping-stones to generate a future with promise. All who endure the journey will find God waiting with open arms to welcome each one into the blessings of the afterlife.

> Troubled days follow human birth,
>
> As a suckling reaches Mother earth.
>
> A stumbling earthling emerges to life
>
> Bursting forth with a budding delight.
>
> Walking with a stumbling gait
>
> Trouble lurks at the garden gate.
>
> Youth blossoms in the morning light
>
> Withering age is reaped before night.
>
> The frailty of life takes a final nod
>
> Reaching for the hand of a loving God.

(Rendering of Job 14:1-2 EDOT)

Most reforms in faith-based groups were to ***"correct the inferior and reconstruct the superior"*** not to start a new name brand expression of faith. The effort was normally to restore scriptural teaching rather than follow the rules of men. Luther's purpose was to repair and rebuild morality on a scriptural foundation rather than on the efficacy of indulgences and the false pardon of sinful behavior. As an academically prepared professor, he proposed a change that was unacceptable to some. The Wesley brothers did not leave the Church of England, they initiated a method of prayer and taught faith-based living. When Charles Wesley died there were at least 500 ordained ministers of Methodism and they broke from the Church of England. The Restoration Movement in America had the objective to restore New Testament teachings but became a series of sectarian groups. The Holiness/Pentecostal movement began as a Gospel Union to restore lost elements of the New Testament. Even Islam began as an effort to free the world from polytheism and focus on Abrahamic Faith in the One True God Creator and Sustainer of the Universe. Early Islamic leaders believed that because Judaism and Christianity had failed to advance morality and the One God of Abraham and this was the reason that moral corruption was rampant in the world. Perhaps men still get in the way of spiritual renewal.

A prefix of *"pro"* or *"con"* are converse or opposite each other; *"pro"* used as a prefix means advancing or projecting forward and when occurring on words from Latin means "forward." The concept of protestant originates from the Latin word *"protestari"* meaning *"declare publicly, testify **for** some-thing."* The state of prolonged dispute over conflicting perspectives is a *controversy.* Protest was coined from two Latin words meaning *"turned in an opposite direction"* or *"to turn against."* A pro-test requires a positive assessment of the situation and an agenda for constructive change. A classic protest should be viewed as an *"objective complaint or an affirmation of a different position."* All civil disobedience must be factually based and structured toward a constructive conclusion with the understanding that *"slinging mud is actually losing ground"* for the cause at hand. An early truism in philosophy was clear: **"one never reaches a positive conclusion beginning with a negative premise."**

Whether reform came from Martin Luther or Dr. King, both being fervent leaders, their model was non-violent and established a pattern for constructive change. When Luther nailed his 95-theses to the church door in Germany, his effort was to bring behavioral change to the spiritual life of the German people. He used the proposition, *"the just shall live by faith"* as an optimistic and constructive strategy to gain affirmational change. Luther delineated specific entities that were founded in faith and the items that should be changed to be in line with scripture. On the other hand, Dr. King was clear that he dreamed of a better day in the American civil society and that his assessment against injustice and discrimination were not in line with the governmental documents: *"We hold these truths to be self-evident: that all men are created equal, that they are endowed by their Creator with certain unalienable rights, that among these are life, liberty, and the pursuit of happiness."* In Jefferson's draft, these words were *"sacred and un-*

deniable," but Franklin edited the draft to read "self-evident" and it was approved by vote. This is the way true statesmen solve differences – through the voting process not public rebuke or denunciation. All constructive protest must be assessed, evaluated, legal, non-violent and led by mature and wise elected officials trusted by the people.

Perhaps present-day protesters fail to understand the ethics or the philosophy of constructive social change. What one sees as positive another may view as negative. Constructive change carries with it an aspect of benevolent humanitarianism, where civil society has a better balance between the *"haves and have nots."* Leadership must control the process; and clear knowledge of how to reform the system or improve a situation must be understood. Just as Luther emphasized the spiritual foundation of *"The just shall live by faith,"* Dr. King advanced moral, ethical, and Constitutional grounds. Surely, honest God-fearing people can find a moral basis for constructive change.

Faith-based foundations were used in the same way the Reformers of the Middle Ages emphasized the authority of the "preserved scripture;" it was above the impulses, frenzies, and rages of a few. Even perceived injustice or disagreement requires clarity and understanding rather than destructive controversy. Any and all resistance to policy, personalities, or parties should be based on a valid argument from the founding principles that were preserved in official documents and actions. Otherwise, we have lawlessness, riot, and disorder. Built into Constitutional Government are ways and means to present logical change when there is injustice or oppressive government pressure to behave against founding principles or the best practices of business and industry.

> *Human beings we may be, but we do not fight our battles in human strength: 4. the weapons we use to fight are not human, but mighty through God to demolish strongholds; 5. casting down the conceits of men against the knowledge of God and bringing every human thought into obedience to Christ; 6. when your submission reaches completion, I am prepared to settle the scorecard with the disobedient.* (2 Corinthians 10:3-6 EDNT)

WEAR THE COMPLETE ARMOR OF GOD*

10. Finally, my brothers, be strengthened in the Lord, and in the power of His unlimited resource. 11. Wear the complete armor of God, so you can stand against the strategy and assault of the adversary. 12. For our wrestling is not against a physical enemy, but against evil princes of darkness who rule this world, against hosts of spiritual wickedness in heavenly warfare. 13. Wherefore wear the complete armor of God that you may be able to withstand evil attacks when they come and be found still standing. 14. Stand your ground, being protected by Truth, and having integrity for a breastplate; 15. and the gospel of peace preparing

*Artwork by Brian Lane Green

your feet for battle, 16. above all, take the shield of faith to extinguish all the fiery darts of the wicked. 17. And take the helmet, which is salvation, and the sword of the Spirit, which is the word of God: 18. praying on every occasion through petition in the Spirit and be vigilant with unwearied perseverance and supplication for all saints; 19. and pray for me, that fluency of speech may be given me, that I may make known courageously the sacred secret of the gospel, 20. for which I am an envoy in a coupling-chain bound to a guard: that in spite of that detail I may speak bravely, as I ought to speak. (Ephesians 6:10-20 EDNT)

... practice these virtues, you will make no false steps: and you shall be supplied the entrance into the kingdom.

2 Peter 1:8-11

ARE ALL BELIEVERS EQUAL?

THE POWER OF MORAL EXCELLENCE

Step XII

Refresh Your Memory of These Things

(⊕) *Internal direct sum or memory*

Refresh Your Memory –2 Peter 1:12-16

12. It is for these reasons that I intend to constantly remind you of these things, although you know them well, and are grounded firmly in your memory; 14. the Lord Jesus Christ has showed me that shortly I must fold my tent. 15. Moreover, I will make it my endeavor that after my departure you will always remember these things. 16. For we have not pursued deceitfully devised folktales, but were eyewitnesses to His majesty when we made known to you the power and presence of our Lord Jesus Christ

Peter wrote his second Letter to believers about AD 67. The purpose was to refresh their memory and remind them of essential things: an exhortation to spiritual growth; the necessity of holding on to truth; warnings against false teachers; and advice on lifestyle in view of the Lord's return.

Peter had confidence that those who were fully taught would remain strong in the faith by recalling those things grounded in their long-term memory and woven into the fabric of their faith. He was concerned that the cares of life and human distractions would hinder the operation of learned truths about the power and presence of the Redeemer in their daily life. Peter was concerned about their future lifestyle. He not only believed they were grounded in the faith, but that they had the capacity to "remain" faithful and maintain an excellent moral lifestyle.

What were "those things" Peter wanted the reader to remember? It must have been the steps from conversion to moral excellence expressed in the early part of the letter. Let us refresh our own memory of "these things" related to their common faith and precious promises: benevolent love, brotherly kindness, godly worship, steadfastness, self-control, learn from books and teachers and reach the level of moral excellence. Peter wanted to refresh their memory that God's compassionate love was the foundation of their faith and that His promises were "value added" to enable their currency in their missional lifestyle. In his reminder, Peter was careful to remind the believers that God's steps to a moral lifestyle were spiritual guidance growing out of God's benevolent love. Moral excellence comes from books and teachers and must be systematically taught and exhibited by mature believers.

What do you remember about your early childhood? Perhaps the song ***"Jesus Loves Me."***

Jesus loves me, this I know
For the Bible tells me so.
Little ones to Him belong
They are weak, but He is Strong!
Yes, Jesus loves me, this I know
For my Mother told me so.

"Remember "These Things"

7. Moral Excellence

6. Knowledge from Books and Teachers

5. Self-control

4. Enduring Steadfastness

3. Godly Worship

2. Brotherly Kindness

1. Benevolent Love

Early on, basic education included "4Rs"-reading, 'riting, 'rithmetic (and a little religion). This was years before language philosophy became critical in subject matter sharing. Words, symbols and numbers became primary elements of communication theory. This included the conveying of ideas, concepts, constructs, data and information and created a common part of the teaching-learning process which enabled integration and interpretation of language. This initiated conceptual systems of words, symbols, or numbers including the sophisticated process of computer computations. The computer system gave immediate value to social scientific statistics and opened the door for more effective social research on matters of interest to individuals, families, communities and faith-based operations.

Early on, "number theory" was considered useless, yet it became essential to the development of computer encryption systems. There are other uses of Applied Mathematics involved in problem solving. Math is not a science; the sciences view data in various areas and develop new or improved methods to meet current and future challenges that may not be connected to observable human behavior. Mathematical logic explores the ability to reason, construct problem-solving mechanisms*, useful operations and applications of formal logic to the foundations of philosophy and theoretical knowledge. When phenomenon-based learning is used in serious study, it is a multi-disciplinary, constructive form of reasoning and learning where a subject or concept is viewed holistically rather than with a subject-based restriction. Why not use Applied Mathematics to explain certain common aspects of real-life in areas of morality, ethics and faith-based behavior?

> *Mechanism is instruction that natural processes (such as life) may be mechanically determined and capable of complete explanation by the laws of physics and chemistry.

1. **Memorization** is being able to repeat word for word what the textbook or the teacher said and is the first step in learning.
2. **Understanding** what the words mean increases but does not complete the required comprehension.
3. **Expressing** the meaning in your own words becomes evidence of understanding.

4. **Providing evidence** that the understanding and expression of the meaning is supported by other sources is an essential step in learning.
5. **Applying remembered information** to answer questions or solve problems formulates usable knowledge and places the data in the long-term memory for easy recall.

A simple restatement of the five (5) steps:
1. **What** does the lesson say?
2. **What** does the lesson mean?
3. **How** could my understanding be expressed in my own words?
4. **Where** do I find collateral evidence to support my expressed understanding?
5. **How** may this knowledge be used in real life?

All education is about the future. Learning is the means to an end and not an end of itself. Everything learned prepares one for the future. This is why the process must proceed step by step until what is learned is used to answer questions or solve problems. Recalling data, information, facts from the memory bank is a good thing, but this is only the first step toward understanding and formulating knowledge to be stored in the long-term memory. Remembering the five (5) steps in the learning process which develops usable knowledge could benefit the process.

Memorization is data to answer, a question or to solve a problem. At this point learned facts becomes knowledge to be used in the future. Consequently, a recall mechanism must be developed that provides easy access to learned facts. This is why the wisdom of sacred writings must be diligently searched for truths that provide answers and ways and means of solving personal and social problems in a constructive manner. We may never gain the wisdom of Solomon, but we can learn from his words:

> *Two are better than one; because they have a good reward for their labour. For if they fall, the one will lift up his fellow: but woe to him that is alone when he falleth; for he hath not another to help him up. Again, if two lie together, then they have heat: but how can one be warm alone? And if one prevails against him, two shall withstand him; and a threefold cord is not quickly broken.*

(Ecclesiastes 4:9-12 KJV)

The transfer of knowledge is a concept where something learned in one field of study may be useful in another area. Learned professionals should always be aware of what they learn in one academic field may assist their functional reality in a faith-based operation. Public education and in particular higher levels of learning were broadly based or generalized as to what the student may need in the future. Then specialization narrowed educational offerings until a student was forced to choose a career direction before they were sufficiently matured to commit to a select direction for their future.

> **Sidebar:** A three (3) hour vocational aptitude evaluation was given to students. Thirty days later a Professor from New York came to interview and discuss the evaluation with those who took the exam. When my turn came, the Professor said, *"Mr. Green, according to your answers on the evaluation you should be a businessman, a politician, or consider a career in the military."* Surprised, even shocked, my mumbled response, *"My mother is a teacher, and my father was a Church Deacon, since age eleven my direction has been in some area of 'ministry through*

education." The Professor's response, *"That choice would be fine! A clergy/educator needs the qualities of a businessman, the vision and personality of a politician, and the leadership of a military officer."*

This book is a reminder to leadership to remember that all truth is of God and should never be ignored under any circumstance. God provided sacred scripture and gifted authors and teachers to facilitate learning. In addition to God's sacred toolbox, many tools and mechanisms exist outside the domain of religion that could assist with the difficulties that exist in the faith-based arena, but they have to be understood and adapted to a secondary use to become an instrument of value. Faith-based leadership must think outside the box.

Refresh your memory of all your past learning regardless of the source. There are nuggets of truth hidden in the lessons, class interaction, exams, developmental readings, or research findings that could be a "value added" asset to those who will recall, review, refresh, and resolve to find ways to apply all truth regardless of how God made it available.

Perhaps we overlook the value of words from a high tempered old fisherman who in a rage cut off the ear of a man seeking to harm Jesus. Remember, Peter was spiritually restored to a position of leadership and obviously used both books and teachers to learn more about his calling and ministry opportunities. In Peter's list of steps to an excellent moral lifestyle, the final step was **"knowledge from books and teachers."**

Teaching is a worthy profession. It should be remembered that all leaders in the New Testament were required to be "apt to teach." This capability came from advance preparation and having "a teachable spirit" themselves. A teacher cannot teach what they do not know. A student cannot learn without advance preparation, attendance with interest in the subject at hand, a willingness to do assignments and use a career or calling as an applications laboratory.

More is learned and education becomes an asset when good teachers are remembered. My fifth-grade teacher, Mrs. Hoodenpyle, stirred in me the desire to learn, to be a reader of books, to understand the meaning of words, and to be a life-long learner. She spoke a private word to me one day as we left the classroom. *"Learning does not stop at the classroom door; it continues every day throughout your whole life."* Such teachers should be remembered, and their example followed.

What is an interdisciplinary education? This is the ability to view a subject or phenomenon through an integration of several disciplines and includes an effort to overcome prejudice and discrimination that improves mutual understanding among peoples and cultural groups. The best advice about such personal development and education came from Col. Creed Bates, Retired, U.S. Army. Learning that I had decided not to make Military Service my career, Col. Bates called me to his office, I saluted, reported and was asked to be seated. Looking straight at me, Col. Bates said, *"Don't think you can go to school for 4, 5, or 6 years and ever communicate with anyone. What you need to do is go to school part-time and work full time with people. You must stay in touch with the real world; not get lost in academia."* Out of this advice came my interest in interdisciplinary education with an alternative delivery system for mature students living in the real world of family and work difficulties. The alternative delivery system became a both/and semester or term system with class-based credits in residency and field-based research with a career or professional applications laboratory, and the writing of books to reach different levels of audience.

PART FIVE: Back Material

AFTERWORD
BEWARE OF LEAPFROG DIDACTIC SCHEMES (≠)

ABOUT THE AUTHOR

APPENDIX A
OTHER BOOKS BY THE AUTHOR

APPENDIX B
GOD IS!

APPENDIX C
HOW GOD COMMUNICATES

APPENDIX D
MENTORS AND COACHES

APPENDIX E
GUIDANCE FOR CONVERTS

BIBLIOGRAPHY

(≠)

Afterword

Beware of Leapfrog Didactic Schemes

(≠) *Symbol for not equal*

1. There were false prophets also among the people, and there will be false teachers among you, who secretly will bring destructive opinions among you, even denying the Master who bought them, and they will bring swift self-destruction. 2. Many will embrace their unashamed immorality and through them the True Way will be brought into disrepute. (2 Peter 2:1-2 EDNT)

Believers must be aware of destructive opinions among the people that are contrary to the True Gospel. Just as individuals attempt to find short-cuts to heaven, some faith-based leaders promote *"easy believe-ism"* which permits sin filled lives to leapfrog over genuine conversion and falsely believe they can walk on streets of gold without reference to a moral lifestyle on earth.

Didactic schemes include efforts to teach what is designed to provide faith-based instructions pleasing to the hearer and supports "easy believe-ism." Yes, God forgives past wrongdoing, and parents, the extended family, teachers and faith-based workers are to deal with present behavior that makes individuals moral citizens of society. However, it takes not only a divine encounter but a firm spiritual connection to overcome behavioral weakness and transform an earthling into a mystical citizen of heaven.

Faith-based families and friends through systematic guidance and discipline can *"bend the willow while it is young"* and create an atmosphere conducive to repentance and a moral lifestyle. However, using leapfrog schemes and feel-good sermons will not accomplish the task of turning converts into learners and fulfill Jesus' Challenge "as you go into all the world, make disciples, identify them with the work of the Trinity through baptism and teach them to observe all things whatever I have commanded you." Jesus concluded that He would be with believers: present now and ALWAYS, even to the end of the journey. (Matthew 28:16-20)

> *17. I appeal to you brethren, make record of all who cause divisions and transgressions against the doctrine that you learned, and avoid them. 18. For such do not serve the Lord Jesus Christ, but are slaves to their base desires, and by pleasing words and smooth speeches deceive the hearts of the innocent.* (Romans 16:17-18 EDNT)

The 24-7 news with biased commentary and no grounded facts, most are aware of the obvious bias expressed in the opinions of those involved in the public press. Parents ought to be aware of what their children see on the internet, watch on TV, and what is taught in public schools and colleges. It is obvious, with recent public exposure of clergy misconduct that was overlooked, by those who had oversight. Should an immoral leader teach morality? Should a pedophile be trusted with children? Must we continue to elect problematic leaders to make laws that they themselves break on a regular basis? Should stakeholders allow morally weak leaders to remain in office only because they are rich? Should faith-based people participate in corporations or congregations led by morally weak people?

Compassionate caring and generous affection are hallmarks of believers, with a positive attitude about the future for each individual in their sphere of influence. Life is not a game or play time leapfrog for the amusement of children. Notwithstanding, the limited value of sports and games (1 Timothy 4:8-10), some churchgoing folk play games and provide entertainment for the young instead of following God's Plan. What was the wise plan presented in Proverbs? Note the emphasis on the male child being prepared to support the family materially and guide the family spiritually.

> "Start a male child on the right pathway, and even when he grows whiskers he will remember and not stumble or side-step from the way."

(Proverbs 22:6 EDOT)

When adults should be nurturing children in the fear and admonition of divine accountability, the easy road is often taken. Without accountability there is no responsibility. Children need to be accountable not only to God, but to parents, grandparents, aunts, uncles, cousins, teachers and friends. A perpetual recess with extended "play-time" is not the way to prepare children for adult life. Recreation has short-term benefits while a faith-based moral lifestyle has lifelong and eternal value.

> **Sidebar:** Although gender equality existed in Judaism, the gender roles were different. Women received religious training, but men also received a secular education to enable better care for the family. However, in sacred documents Jewish men were given religious importance because only males could become a Rabbi. In the Hebrew culture both

genders provided spiritual guidance and discipline for the children, but men had the primary role of moral example and providing for the basic needs of the family.

What young adults recall from their childhood, it will not be the rules of leapfrog or the game of dodge ball. If you ask any adult about who influenced them during childhood, it will be the older adults from their extended family, their early teachers, or an elderly couple in the church or community. What did they remember? Practical things about life and living learned either by observation, imitation, or direct instructions. Adults must weigh the relative value of juvenile games vs spiritual guidance in allotting time, talent, and guidance in preparing the young for a moral and ethical lifestyle. It is called *"tough love."*

> 7. Be patient, then, for while correction lasts; God is treating you as His children. Was there ever a son whom the father did not correct? 8. If you are left without discipline, that discipline which everyone must share, then you are illegitimate children and not sons. 9. We have known what it was to accept correction from earthly fathers, and with reverence, shall we not willingly line up under the authority of our spiritual Father, and live? 10. It was for a short time that our earthly fathers disciplined us as they thought best; but God disciplines us for our highest good, to give us a share in His holiness. 11. At the time all discipline is painful rather than pleasant; but afterwards, when it has done the work of correction, it yields a harvest of good fruit in a righteous lifestyle for those trained by the experience. 12. So, then, lift up the drooping hands, and the weak knees; 13. and plant your feet in a straight path, lest

someone who is weak stumble out of the path; but be restored to strength instead.

(Hebrews 12:7-13 EDNT)

Some faith-based folks are so heavenly minded they have little earthly influence. They follow man-made rules and are so worldly they stumble on the path that leads to moral excellence. They cannot see the basic needs of others. They are so deeply engrossed in the concepts and constructs of their sectarian position they take their eyes off the scripture and their teaching becomes a scam. They feel no personal responsibility to assist the poor, teach the children, or guide seekers into sacred truth. Such false teachers become *"good bad examples"* and lead others astray. If the blind were led by the blind, they both end up in the ditch of despair. Most individuals desire a better afterlife but are not guided to take the steps to develop a lifestyle worthy of becoming a moral citizen of society as the first steps toward heaven.

Others think *"Let the young get into trouble and then they will seek God."* This approach delays the process of positive decision making with reference to redemption. At the cross there were two thieves, one cursed Jesus and the other embraced His love and asked, *"Remember me when you come into your Kingdom."* This record of *"just in time"* faith-based conviction was so no one would despair, but only one so no one would presume on the mercy of God. Faith which leads to moral living is the beginning work of redemptive grace. *"For just as the body separated from the spirit is lifeless, so faith without works is already a corpse."* (James 2:26 EDNT)

All aspects of life and living have to do with the future: the past cannot be changed only forgiven, the present can be improved with mature guidance, but the future may be changed only through faith and spiritual guidance from books and teachers. This aspect of human development and surrogate parenting should mature into *"benevolent humanitarianism."* The poor and the young are eager for good news about faith and future things; however, faith that is only words without the required works is deadly. To experience God's love, one needs a human connection to faith. The naked must be clothed, the hungry must be fed, and inquiring minds must be filled with sacred wisdom before they truly listen to the message of grace and accept the struggle between their humanity and divine influence on their life.

Most families and faith-based trail blazers get things out of order by delaying the process of spiritual development. This permits the young to grow up too fast without learning basic facts for their spiritual formation. Bringing up a child to be a moral citizen of society is a team effort. Parents, family, teachers and faith-based workers all have a role in this progression. Growth and maturity are parts of an incremental process. All must learn to believe, learn to obey, learn to trust, learn to love, learn to forgive, learn to share, and learn the value of God's Word and spiritual friends. This process requires patience, love, and good examples of a productive and moral life. The young learn to bond, develop personality, build a knowledge base, construct character, and form their faith-based foundation prior to age 12. As teenagers they listen only to peers, public school teachers, and secular media and begin to spread their wings without the moral roots normally supplied by family and early friends. This roadway is full of moral potholes and dark and dangerous places. And each journey around the sun brings more despair. Drifting along like a tumble weed is not the way to grow strong and prepare for the future.

Timely and loving discipline by parents is essential for a good foundation for life. Without firm grounding, the future has many dangers. Without roots there will be no wings to rise above the storms of life. Basic human flaws must be corrected before a moral lifestyle can be constructed to support a meaningful purpose for living. Missed opportunities in discipline create an obligation for additional future correction. This further complicates the maturing process and constructive decision making at the age of accountability. The young must be prepared to make "value added" decisions when they begin to take responsibility for their behavior. When this point is reached, a faith-based acceptance of a changed life will save not only one life, but perhaps a future family, and many friends and coworkers who recognize a moral lifestyle in contrast to their own. This is the ultimate spiritual witness.

Maturity and spiritual refinement are not gained instantly or without growing pains. It comes from *"precept upon precept and line upon line."* A precept is a *"guiding principle to influence behavior."* The Ten Commandments are examples of early precepts used to guide people to the Promise Land. Such principles must guide the young into adulthood and productive moral citizenship.

> **1. But you must speak those things that are appropriate for healthy teaching:** 2. Charge the senior men to be sober, serious, prudent, healthy in Christian faith and love and endurance. 3. The

senior women likewise must behave appropriately for a holy calling, not given to slanderous talk or given to wine, teaching others by good example: 4. in order that they may train the young women to be lovers of their husband and child lovers, 5. to be sensible, pure, homemakers, good, lining up under the authority of their husbands lest the word of God be abused with foul language. 6. The younger men similarly exhort to be sensible, **7. about all things, showing yourself a pattern of good works in your teaching display purity of motive and seriousness. 8. Present a wholesome message that cannot be criticized;** *in this way your opponents may be ashamed, having nothing disparaging to say to you.* (Titus 2:1-10 EDNT)

The young must grow in grace and knowledge over time. Children should be taught moral principles and human kindness in the process of building character and early spiritual formation prior to the age of accountability. When this progression occurs, the pathway toward becoming a moral citizen of society becomes clear. Yes, God desires that all be saved and brought into His Kingdom. Providence takes care of the early years of innocence, but parents, family, teachers and spiritual leaders must not abandon their systematic guidance and spiritual coaching to prepare a child for responsibility as an adult. The longer the delay past the age of accountability the young are confronted with a life change encounter, the process becomes increasingly more difficult.

One does not climb a mountain starting at the top, or climb a ladder starting in the middle. The first steps must be carefully negotiated. Human nature and nurturing of the young require early faith-based steps in development.

Children are not taught to add a column of figures starting in the middle or at the top. Starting with the foundation stones and move step by step upward is the normal process. Family members often permit a child to be "little adults" by omitting basic steps in the growing process. This is probably the greatest hindrance to mature development and spiritual formation. Normally, one is taught the essential elements of a subject before moving to more complex subject matter.

> **Sidebar:** Note an example in Luke 10:25-37. A certain man had memorized spiritual material without understanding or learning to believe and behave what had been learned. This made the process impossible to complete because he did not recognize his responsibility toward others. **Yes, he memorized scripture, but it was not woven into the fabric of his faith.** He was preoccupied with desire for eternal life instead of understanding that he must believe and behave the Word and understand the process of reaching moral excellence. Someone had failed to complete the task of preparing him to make a constructive decision when given the opportunity. *Not only a positive decision that launched him on a plan to construct a moral lifestyle.*

10. But you have closely followed my teaching, the conduct, the purpose, the faith, the long-suffering, the love, the endurance, 11. the persecutions, the sufferings, which happened to me in Antioch, Iconium, and Lystra; which persecutions I endured: but the Lord delivered me out of them all. 12. Yes, all who will live godly lives in Christ Jesus will be persecuted. 13. <u>But wicked men and imposters will keep on going from bad to worse, misleading others and deceiving themselves.</u> **14. But continue to hold fast the things you have learned and been convinced of, knowing the teachers from** *whom you learned them; 15. and from early childhood, you have known the sacred letters, the ones able to make you wise unto salvation through faith in Christ Jesus.* 16. All sacred writings are God-breathed, and serviceable for teaching, for warning, for correction, for instruction in righteousness, 17. in order that the people of God may be adequately equipped for every good work. (2 Timothy 3:10-17 EDNT)

ABOUT THE AUTHOR

Hollis L. Green, ThD, PhD, DLitt, is a Clergy-Educator with public relations and business credentials and doctorates in theology, philosophy, and education. A  Distinguished Professor of Education and Social Change at the graduate level for over four decades, Dr. Green is a Diplomate in the Oxford Society of Scholars, and author of 50+ books and numerous articles. He served six years as a member of the U.S. Senate Business Advisory Board and maintained certified membership in several public relations societies (RPRC, PRSA, and IPRC). He served pastorates in five states, a Military Chaplain during the Vietnam era, a denominational official for 18 years, and traveled in ministry and lectured in over 100 countries.

Dr. Green was the founder (1974) A.I.D. Ltd., Associated Institutional Developers, Ltd. (an international Public Relations and Corporate Consultant Company. He was Vice-President (1974-1979) of Luther Rice Seminary (www.lrs.edu) and became the founding President (1981) and Chancellor (1991-2008) of Oxford Graduate School. He continues to serve this institution as Chancellor Emeritus [www.ogs.edu]. As part of a global outreach, Dr. Green founded (2002)

Oasis University in Trinidad, W.I. [www.oasisgradedu.org] where he continues to serve as a Professor of Education and Social Change and Chancellor. In 2004, he assisted in establishing Greenleaf Educational Foundation in Colorado to advance higher education.

In addition to other endeavors, Dr. Green launched Global Educational Advance, Inc. and GEA Press (2007) [www.gea-books.com] to advance higher education and constructive social change through publishing, curriculum advance, library/learning resources, improved instruction, and global book distribution. His books and assisting authors in publishing are a logical outgrowth of a sixty-year ministry through education. He serves Global as Corporate Chair and Co-publisher with his sons, Barton and Brian. Dr. Green continues to speak, teach, write books and work with authors in publishing quality creative work. He also maintains an interest in Childcare and Eldercare facilities, and Oxonian Learning Centers globally.

APPENDIX A

Recent Books by the Author

To understand the problems of faith-based entities and advance delivery systems for graduate education, extensive research was done during the past four decades. Meanwhile, Dr. Green's schedule was filled with academic administration, teaching, research and writing, but colleagues and friends have encouraged sequels to his best-known works. In two decades following retirement, Dr. Green followed that prompt and produced twenty (25+) books plus ten Children's Novellas.

(See *gea-books.com/bookstore*) or anywhere good books are sold including the Expresso Book Machine © worldwide.

Why Churches Die. (2007) ISBN 978-1-9796019-03 A fresh assessment of congregational vitality to determine thirty-five reasons why faith-based congregations were losing their pristine power of outreach.

Interpreting an Author's Words. (2008) ISBN 978-0980-164-74—Define both formal and informal study and writing skills by understanding how to clearly interpret the spoken and written words of others.

Discipleship. (2010) ISBN 978-0-9796019-5-8--A revived edition to better explain the process of a believer's lifestyle from conversion (change direction), to discipleship (learning), to apostle (mature enough to be trusted with the message grace.)

Sympathetic Leadership Cybernetics. (2010) ISBN 978-1-9354345-28 – This work attempts to clarify management and leadership in the context of organizational and institutional functionality and charts a course for organizations to serve the needs of people through shepherd management and servant leadership.

Why Christianity Fails in America. (2010) ISBN 978-0-9796019-10-- A call for an internal redirection of the heart and soul to make the pristine faith viable in the Twenty-first century.

How to Build a Better Spouse Trap. (2010) ISBN 978-1-9354344-50 – A major failure of faith-based groups is they have made little difference in the lives of individuals and their function in the family. How to choose a mate, learn for our mistakes, stay married, and teach others to break the cycle of dysfunctional relationships. The family unit is a microcosm of faith-based behavior.

SO TALES. (2011) ISBN 978-1-9354345-80 -- Preserving 240 true stories from the past for the benefit of family and friends.

Designing Valid Research. (2011) ISBN 978-1-9354345-73 – A guide to designing a research proposal and developing a social scientific dissertation.

Titanic Lessons. (2012) ISBN 978-0-9796019-6-5 – An effort to demonstrate that bigger is not necessarily better and that all building of machines, organizations, and institutions must use material that meets the precise requirements of the task. This must be applied to people, process, and functionality of the human element and the mechanics must match the environment.

Why Wait Till Sunday? (2012) ISBN 978-1-935434-27-6 – A renewal plan for older congregations who depended on programs coming down from sectarian authority rather than locally generated ideas and involvement in seven (7) aspects of renewal.

Fighting the Amalekites. (2013) ISBN 978-1-935434-30-6 – The unhealthy addictions, unproductive habits, an uncontrolled tongue are all little "Amalekites" unless these are destroyed, they will become the destroyer. These join the Amalekites that ambush and take advantage of spiritual weaknesses.

Remedial and Surrogate Parenting. (2013) ISBN 978-1-9354344-81– Children are a gift of God and a legacy of faith-based families; therefore, parenting skills are an essential aspect of religion. This work is guidance for remedial human development (0-20) for parents, teachers, and childcare workers.

Transformational Leadership in Education. (2013) Second Edition ISBN 978-1-9354342-38 – A strengths-based approach to education for administrators, teachers, and guidance counselors.

Tear Down These Walls. (2013) ISBN 978-1-9354341-84 – A priority agenda must be to make people moral citizens of the world before they can become mystical citizens of heaven. Where organized groups choose not to function, personal action could make a difference and break down some of the barriers that divide the faith-based community and strengthen the "One Lord-One Faith–One Baptism" message.

The EVERGREEN Devotional New Testament – C.A.F.E. Edition. (2015, 2019) ISBN 978-1-9354342-69 – EDNT is a 42-year project to translate common NT Greek and determine the meaning "then" and how words can best be expressed "now" and remain true to the original intent expressed in a common devotional language.

Recycled Words n' Stuff. (2016) ISBN 978-1-9354348-63 – A collection of short narratives and essays of general interest.

The Children's Bread – Unlocking Whole Life Stewardship. (2018) ISBN 978-1-935434-90-0 Appreciating faith-based economics and personal wealth to unlock a missional lifestyle and funding for humanitarian and faith-based entities.

Kingdom Growth Through Missional Behavior (2019) ISBN 978-1-935434-91-7—Adopting the thinking, behaviors, and practices of a missionary in order to globalize the message of grace.

God Has Confidence in You. (2020) ISBN 978-1-950839-04-9. No test has come your way, but such as is common to man: God is faithful, who will not permit you to be tested beyond your endurance; but will with each test also how you a way forward, so that you may be victorious.

Power of Forgiveness. (2020) Forgiveness is the Sunrise of Reconciliation. ISBN 9781950839063

Beyond Pulpit, Classroom and Lecture Hall. ISBN 978-1-950839-03-2 (2021) – Unlocking Exposition, Instruction and Research Reporting

Navigating Multiculturalism. (2021) Guidance for Constructive Sociological Change. ISBN 978-0-9796019-4-1

Are All Believers Equal? (2022) –The Power of Moral Excellence (2 Peter 1:1-21) ISBN 9 78-1-950839-10-0 How Applied Mathematics, Linguistics, and Hermeneutics inform the interpretation of language using 2 Peter 1:1-21 to develop twelve steps to moral excellence.

Constructive Social Change Research – The Pursuit of Knowledge (2023) – ISBN 978-1-950839-00-1 Guidance in development of a master's thesis, designing a doctoral research proposal and constructing a defendable dissertation based on social scientific research with an objective of constructive social change.

Plus Children's Books:

In keeping with the Dons of Oxford University, a dozen Children's Novellas available as books or PDF:

Sleepy Town Lullaby and Story 978-0-9796019-4-1

The Scoop about Birthday Soup 978-0-9796019-8-9

- Cranky Not-so-Hottra'
- Cat-Astropic Charlie.
- The Funky Chicken's Wedding.
- A Tea Party at Nany's House.
- The Shimonaka Big Dripper!
- The Mouse of the House.
- The Boy Who Wanted to Grow a Beard.
- The Trouble with Funny Book Cussing.
- The Blue Jay and Grandma's Song.
- Ditala Killed a Dead Snake

APPENDIX B

This side of eternity, it is impossible for our human minds to fully comprehend God's infinite and awe-inspiring nature. In the Bible, however, He has shared enough truths about Himself to draw us into faith and worship.

We encourage you to commit five minutes daily over a 30-day period to prayer over these verses and worship God in new ways. You'll be amazed by how much closer you feel to Him at the end.

30 DAYS OF PRAYING THE

Names and Attributes of God

JEHOVAH

The name of the independent, self-complete being—"I AM WHO I AM"—only belongs to Jehovah God. Our proper response to Him is to fall down in fear and awe of the One who possesses all authority.
Exodus 3:13-15

JEHOVAH-M'KADDESH

This name means "the God who sanctifies." A God separate from all that is evil requires that the people who follow Him be cleansed from all evil.
Leviticus 20:7,8

INFINITE

God is beyond measurement— we cannot define Him by size or amount. He has no beginning, no end, and no limits. Romans 11:33

OMNIPOTENT

God is all-powerful. He spoke all things into being, and all things—every cell, every breath, every thought—are sustained by Him. Nothing is too difficult for Him. Jeremiah 32:17,18, 26,27

GOOD

God is the embodiment of perfect goodness, and is kind, benevolent, and full of good will toward all creation. Psalm 119:65-72

LOVE

God's love is so great that He gave His only Son to bring us into fellowship with Him. His love encompasses the world, and embraces each of us personally and intimately. 1 John 4:7-10

JEHOVAH-JIREH

"The God who provides." Just as He provided yesterday, He will provide today and tomorrow. He grants deliverance from sin, the oil of joy for the ashes of sorrow, and eternal citizenship in His Kingdom for all those adopted into His household. Genesis 22:9-14

JEHOVAH-SHALOM

"The God of peace." We are meant to know the fullness of God's perfect peace, His "shalom." God's peace surpasses understanding and sustains us through difficult times. It's the product of fully being what we were created to be. Judges 6:16-24

IMMUTABLE

All that God is, He has always been. All that He has been and is, He will ever be. He is ever perfect and unchanging. Psalm 102:25-28

TRANSCENDENT

God is not simply the highest in an order of beings (this would be to grant Him eminence). He is transcendent—existing beyond and above the created universe. Psalm 113:4,5

JUST

God is righteous and holy, fair and equitable in all things. We can trust Him to always do what is right. Psalm 75:1-7

HOLY

God's holiness is not a better version of the best we know. God is utterly and supremely untainted. His holiness stands apart—unique and incomprehensible. Revelation 4:8-11

JEHOVAH-ROPHE

"Jehovah heals." God alone provides the remedy for mankind's brokenness through His son, Jesus Christ. The Gospel is the physical, moral, and spiritual remedy for all people. Exodus 15:22-26

SELF-SUFFICIENT

All things are God's to give, and all that is given is given by Him. He can receive nothing that He has not already given us. Acts 17:24-28

OMNISCIENT

God is all-knowing. God's knowledge encompasses every possible thing that exists, has ever existed, or will ever exist. Nothing is a mystery to Him. Psalm 139:1-6

OMNIPRESENT

God is everywhere, in and around everything, close to everyone. "'Do not I fill heaven and earth?' declares the Lord." Psalm 139:7-12

MERCIFUL

God's merciful compassion is infinite and inexhaustible. Through Christ, He took the judgment that was rightfully ours and placed it on His own shoulders. He waits and works now for all people to turn to Him and to live under His justification. Deuteronomy 4:29-31

SOVEREIGN

God presides over every event, great or small, and He is in control of us. To be sovereign, He must be all-knowing and all-powerful, and by His sovereignty He rules His entire creation. 1 Chronicles 29:11-13

JEHOVAH-NISSI

"God our banner." Under His banner we go from triumph to triumph and say, "Thanks be to God, who gives us the victory through our Lord Jesus Christ" (1 Corinthians 15:57). Exodus 17:8-15

WISE

All God's acts are accomplished through His infinite wisdom. He always acts for our good, which is to conform us to Christ. Our good and His glory are inextricably bound together. Proverbs 3:19,20

FAITHFUL

Out of His faithfulness God honors His covenants and fulfills His promises. Our hope for the future rests upon God's faithfulness. Psalm 89:1-8

WRATHFUL

Unlike human anger, God's wrath is never capricious, self-indulgent, or irritable. It is the right and necessary reaction to objective moral evil.
Nahum 1:2-8

FULL OF GRACE

Grace is God's good pleasure that moves Him to grant merit where it is undeserved and to forgive debt that cannot be repaid.
Ephesians 1:5-8

OUR COMFORTER

Jesus called the Holy Spirit the "Comforter," and the apostle Paul writes that the Lord is "the God of all comfort." 2 Corinthians 1:3,4

EL-SHADDAI

"God Almighty," the God who is all-sufficient and all-bountiful, the source of all blessings.
Genesis 49:22-26

FATHER

Jesus taught us to pray, "Our Father" (Matthew 6:9), and the Spirit of God taught us to cry, "Abba, Father," an intimate Aramaic term similar to "Daddy." The Creator of the universe cares for each one of us.
Romans 8:15-17

THE CHURCH'S HEAD

God the Son, Jesus, is the head of the Church. As the head, the part of the body that sees, hears, thinks, and decides, He gives the orders that the rest of the body lives by.
Ephesians 1:22,23

OUR INTERCESSOR

Knowing our temptations, God the Son intercedes for us. He opens the doors for us to boldly ask God the Father for mercy. Thus, God is both the initiation and conclusion of true prayer. Hebrews 4:14-16

ADONAI

"Master" or "Lord." All God's people ought to acknowledge themselves as His servants, with His right to reign as Lord of our lives.
2 Samuel 7:18-20

ELOHIM

"Strength" or "Power": He is transcendent, mighty and strong. This name displays His supreme power, sovereignty, and faithfulness in His covenant relationship with us.
Genesis 17:7,8

THIS TOOL IS MEANT TO BE SHARED.
To download a copy visit *navigators.org/names-of-God*

Sources: *The Knowledge of the Holy,* **by A.W. Tozer;** *Names of God,* **by Nathan Stone; and** *God of Glory,* **by Kenneth Landon.**

APPENDIX C

How a Personal God Communicates

The One God has shared truths about His Nature and distinguishing characteristics as "Father, Son, and Holy Spirit." In (Genesis 1:26) God speaks, "Let us make man in our image, in our likeness…" God views man as a wholistic entity with all parts intimately interconnected and explainable only by reference to the whole, because the whole is greater than the parts. In geometry, an equilateral triangle has three equal sides. So it is with The Creator God who exists in three coeternal and consubstantial personalities* as Father, Son, and Holy Spirit!

> *Personality is the combination of characteristics or qualities that form a person's distinctive character.

Isaiah, The Prophet -

"For as the heavens are higher than the earth, so are my ways higher than your ways, and my thoughts than your thoughts." (Isaiah 55:9)

Paul's words -

"There is a fathomless depth in God's wisdom and knowledge! His judgments are unsearchable, and His footsteps cannot be tracked! Who can figure out the mind of the Lord? Or who could be His advisor? Who has first given to God that He should pay back again? For God is the source, preserver, and ruler of all things: to God is glory throughout all ages. Amen. (Romans 11:33-36 EDNT)

HOW A PERSONAL GOD COMMUNICATES

GOD communicated with man from Creation to the Birth of Jesus as the FATHER.

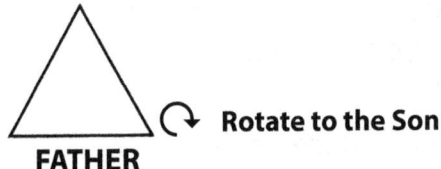

FATHER

GOD communicated with man from the Birth of Jesus to the Ascension as the SON.

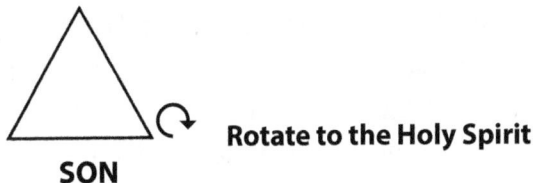

SON

GOD communicated with man from Harvest Festival (Pentecost) to the present as the HOLY SPIRIT.

HOLY SPIRIT ⟶ (∞)

APPENDIX D

Faith-based Mentors and Coaches

When faith-based believers fail to work with new converts both become dysfunctional. Just to welcome them to a good behavior workshop and guide new folk in the way they should go by following Paul's example: "Wherefore, I beseech you, follow my footsteps." (1 Corinthians 4:16 EDNT)

New converts still live in the same family and community. In the past, an application of the essential elements of morality and fairness was left to clergy, judiciary, and law enforcement who have limited contact and little understanding of the realities of the complex community and the dysfunctional families that produce most of the problems of society. There remains a great fixed gulf between the moral convictions and the constructive community agenda. Sadly, the core values and moral fiber of the community have diminished and created a contextual and a lack of justice or fairness for all concerned.

Why does morality and ethics have difficulty in a civil society? Why are the young primary victims of broken promises, broken laws, and broken families? What is the inferior process that weakens the moral fiber? Why has the faith-based message of grace and forgiveness failed to be a viable expression of morality to the community? To integrate basic moral values into the straight and narrow path, one must reach out and find those seeking to change their lives for the better and guide them through conversion and early discipleship to the right way. Priority must be given to new converts, the welfare of the children and the disadvantaged. Hopefully, faith-based groups will have a productive role in this cure and care of souls by consistent discipleship development of converts.

Remedial development and growth begin with confidence. There must be confidence in the watch care provided, confidence in the program and process being utilized as a priority, and also confidence in the personnel and staff with whom one works. Without confidence in these areas, no individual care facility or program can function effectively regardless of the quality or quantity of personnel and funding.

These factors must be considered in the remedial and surrogate nurturing of converts. Understanding how different individuals function in the context of the basic institutions is important to growth and development. Age, background and experience have dissimilar impact on different individuals. All converts and new members of a faith-based group are not equal and must be considered as individuals with both age-level maturity, background, positive and negative experience and personal and emotional feelings. To ignore these aspects is to fail as a disciple maker nurturing new converts in a faith-based lifestyle.

Nurturing a new convert is situational. What may be good care for one person at a particular stage of development may not be good care for another or even for the same person at a different developmental stage. Good care does two things: it matches the person's stage of development, and it empowers the individual to progress toward self-direction. Good care is situational, yet it provides for long-term development by exciting and encouraging converts at the personal level. The entire nurturing system is based on a growth and development from high structure to little or no structure, from a kindergarten-type beginning to work in which the structure or task is lessened over time. The relationship between mentor/coach and the convert or new member continues until the individual is able to function on their own in the real world as a witness to God's saving grace.

Quality and quantity are mutually exclusive; increase one and decrease the other. There must be a proportional balance between these two elements to maintain stable growth and development. There is a limit to what one person can do for another in the spiritual realm. If the staff is short-handed or unskilled, all the new folk undergo some unpleasantness and suffer personal loss. Perfection is not the goal of spiritual care; growth and age-specific development is the true objective. Quality in spiritual watch care comes when new converts and new members each have a guide, coach, counselor, or friend who provide the freedom to search and discover the way forward. Progress in learning should be rewarded with words of praise.

The Critical Path Method (CMP) works backward from a perception of where the person should be at the mature stage. CPM is used in construction when a builder views both an architect drawing of a completed building and a set of working drawings of how the building is to be constructed. The builder, with a clear view of both plans and a conception of the finished product, establishes target dates for each stage of development. The objective is to prepare an individual for the real world when they reach spiritual maturity.

Working back from the target date, the builder considers time, material, and contingencies to establish a construction schedule. At this point the builder must start at the beginning, structure the building in stages, and arrange for an evaluation process based on the architect's plans. Each aspect of the construction must be done in sequence with the timeline affected by the duration of each stage. With the CPM, the builder may evaluate progress using a Performance Evaluation Review Technique (PERT) or an easier method "By their fruit" which some call (BTF). To do this, the mentor/coach must understand clearly the age-specific level of development of each person. Age or physical size is not sufficient criteria to judge progress because each individual is different. Maturity is not measured by age but by attitude, knowledge and behavior. Maturity is a demonstrated movement toward wise and knowledgeable behavior.

The old English "midwife" meant "with woman"; a male who assists with a birth is also a "midwife". What about conversion as a new birth, do we not need some special workers who function as a spiritual birthing agent. Of course, we have to come up with a better name than midwife! What about: believers who are loving mothers and wise fathers. The fruit of the righteous is a tree of life; and he that wins souls is wise. (Proverb 11:30)

> *What does the word that we preach say? It is near you, even in your mouth, and in your heart; 9. if you acknowledge with your mouth the message that Jesus is Lord and believe in saved. 10. For with the heart man believes and is justified; and with the mouth he confirms his salvation. 11. The scripture*

declares that whoever believes on Him shall not be disappointed. (Romans 10:8-11 EDNT)

Provided one follows the path of Jesus from the manger to the Cross, they will see a footpath of self-denial, exhausting travel, and daily service to those in need. Luke wrote, "Jesus healed all in need of healing." Loving service to others was His food and drink and the will of the Father that nourished His earthly life. Consequently, those who follow Jesus will live a sacrificial life demonstrating care and concern for others. There is no room for selfishness or laziness in the life of those who take up their cross and follow the footpath of discipleship. Those who walk in the spirit of Jesus will live a clean life and be doubly blessed. As my friend, Subesh Ramjattan said, "They gather blessings to scatter to others." This is true discipleship and a missional lifestyle.

Not only will committed believers share their God-given resources, they will become personally involved in the work. It is a hands-on lifestyle for them and their family. The harvest fields are ripe and ready for gathering, but not enough labors to complete the task. In 2 Kings 19:3 Hezekiah sent a proverbial message to Isaiah about the divided kingdoms of Israel and Judah being too weak to resist captivity: "We are like a poor traveling women about to collapse without strength to bring their children to birth." This is similar to the circumstance of modern-day faith-based groups. Souls ready for the kingdom, but there is no strength to complete the birthing process and establish true discipleship.

1. As we work together with God, we appeal to you not to accept. the grace of God and let it go to waste. 2. God said, I have heard your prayers at a convenient time, and in the day of salvation I have brought you relief in a difficult situation: observe, now is the time for coming together; now is the day of deliverance. (2 Corinthians 6:1-2 EDNT)

11. The story laid upon me is long and hard to explain, seeing you are so dull of hearing. 12. After all this time you should be teachers, yet you still need to be taught again the first principles of the divine revelation: you have gone back to needing milk instead of solid food. 13. Those who still have milk for their diet do not have the experience to speak of what is right: remains an infant. 14. But grown men can eat solid food, those who, through the development of the right kind of habit, have reached a stage when their perceptions are trained to distinguish between good and evil. 7. You were running the race well; who cut in to obstruct your obedience of the truth? 8. This readiness to believe without evidence does not come from the one who called you. 9. It is true that a little yeast can change the whole batch of dough. 10. I am persuaded in the Lord to have confidence in you, that you will not be led astray: but whoever is shaking your faith will pay the penalty at judgment. (Galatians 5:7-10 EDNT)

Visiting a faith-based service is a stop-gap measure for new converts. They are limited in what they can do daily to cope with their drastic life change. New believers are confronted with dysfunctional or broken families and unkept promises. Yet, faith-based believers have an opportunity to model a missional lifestyle that is observable. The task of mentoring and discipling new believers is the scriptural task of all true disciples of Jesus.

APPENDIX E

Guidance for Converts

- New converts must correct the inferior, before attempting to construct a superior life as a disciple.
- Discipleship cannot grow in the ashes of immorality or develop only as a casual spectator at worship.
- There must be consistent participation in spiritual things to refurbish a Convert with a faith-based Lifestyle!

(1) First Things First

2. since you are newly born, yearn for the unadulterated milk of the word, so you may grow up until your soul thrives in good health. 3. Since you have tasked (tasted) the Lord's kindness. 4. Draw near to him; He is the living fulfillment of the stone discarded by men but chosen of God. 5. You yourselves are lively stones built into a spiritual house; you must be a holy priesthood to offer up spiritual sacrifices acceptable to God by Jesus Christ. (1 Peter 2:2-5 EDNT)

(2) Follow a Good Example

16. Wherefore, I beseech you, follow my footsteps. (1 Corinthians 4:16 EDNT)

(3) Learn from Good and Bad Behavior

19. Now the behavior that belongs to the flesh is obvious, they are: (sensual sins) unfaithfulness in marriage, unrestrained living, unbridled acts of indecency; 20. (religious sins) the worship of idols, the use of drugs and magical powers; (temperamental sins) hostility, strife, jealousy, violent flare-ups of temper, self-seeking ambitions, adherence to contradictory teaching; 21. (personal sins) desires to appropriate what others have, drunkenness and carousing, and similar things: I warned you before that people who do such things will have no part in the kingdom of God. (Galatians 5:18-21 EDNT)

(4) Fruit of the Spirit

22. But the fruit of the Spirit is love, and love brings joy, peace, longsuffering, gentleness, goodness, faith, 23. tolerance and self-control: and no law exists against any of these. 24. And those who belong to Christ have nailed the flesh to the cross with its passions and appetites. 25. Since we live in the Spirit, we should be guided by the Spirit in our orderly walk. 26. Let us not have excessive pride or boastfulness about personal abilities, infuriating one another or causing others to be envious. (Galatians 5:22-26 EDNT)

Bibliography and Reading List

Bicchieri, Constina, (2006) *The Grammar of Society*, Cambridge Press.

Clemson, B. (1991). *Cybernetics: A New Management Tool*. Philadelphia: Gordon and Breach.

Davidson, M. (1996). *The Transformation of Management*. Boston. Butterworth-Heinemann.

Goodman, M. & Karash, R. & Lannon, C. & O'Reilly, K. W., & Seville, D. (1997). *Designing A Systems Thinking Intervention*. Waltham, MA. Pegasus Communications, Inc.

O'Connor, J. (1997). *The Art of Systems Thinking: Essential Skills for Creativity and Problem Solving*. London: Thorsons, An Imprint of HarperCollins Publishers.

Richmond, B. (2001). *An Introduction to Systems Thinking*. Hanover, NH. High Performance Systems.

Warren, K. (2002). *Competitive Strategy Dynamics*. West Sussex, England. John Wiley & Sons.

Babbie, Earl. (2001). *The Practice of Social Research* Wadswoth/Thompson Learning.

Berger, R.M., & Patchner, M.A. (1988). *Planning for Research: A guide for the Helping Professions*. Newbury ParG, CA: Sage.

Behrens, L. (1992). *The American Experience: A Sourcebook for Critical Thinking and Writing*. Boston: Allyn and Bacon.

Bloom B. S. (1956).*Taxonomy of Educational Objectives, Handbook I:The Cognitive Domain*. New York: David McKay Co Inc.

Brockett, R. G. and Hiemstra, R. (1991) *Self-Direction in Adult Learning: Perspectives on Theory, Research, and Practice*, London and New York: Routledge.

Bryman, Alan. (2008). *Social Research Methods*. Oxford University Press.

Candy, Philip C. 1991. *Self-Direction for Lifelong Learning*. San Francisco:

Crosby, B.C. (1999). *Leadership for Global Citizenship: Building Transnational Community.* Thousand Oaks, CA: Sage Publications.

Drucker, F. Peter. (1995). *Managing in a Time of Great Change.* New York: Penguin Group.

Giles, C. (2006). *Transformational Leadership in Challenging Urban Elementary Schools: A Role For Parental Involvement?* University of Buffalo, The State University of New York.

Green, Hollis L. (2007) *Sympathetic Leadership Cybernetics,* Nashville.GlobalEdAdvancePress.

Green, Hollis L. (2007) *Why Christianity Fails in America,* Nashville. GlobalEdAdvancePress.

Green, Hollis Lynn (2008). *Interpreting An Author's Words,* Nashville: GlobalEdAdvancePress.

Green, Hollis L. (2009) .*Remedial and Surrogate Parenting*, Nashille, GloblEdAdvancePress.

Green, Hollis L. (2010). *Designing Valid Research.* Nashville, GlobalEdAdvancePress.

Green, Hollis L. and Swanson, G. A. (2021), *Research Methods for Problem Solvers and Critical Thinkers*, Nashville. GlobalEd AdvancePRESS

Green, Hollis L. (2018) *The Children' Bread – Unlocking Whole Life Stewardship.* Nashville. GlobalEd AdvancePRESS

Green, Hollis L. and Jackson, E. Basil (2019) *Kingdom Growth Through Missional Behavior.* Nashville. GlobalEd AdvancePRESS

Green, Hollis L. (2015, 2019). *The Evergreen Devotional New Testament Complete Edition*, Nashville, Post-Gutenberg Books

Green, Hollis L. (2020) *Power of Forgiveness – Forgiveness is the Sunrise of Reconciliation.* Nashville. GlobalEd AdvancePRESS

Green, Hollis L. (2020). *Beyond Pulpit, Classroom and Lecture Hall – Unlocking Exposition, Instruction and Research Reporting in Subject Matter.* Nashville. GlobalEd AdvancePRESS

Green, Hollis L. (2021). *Navigating Multiculturalism – Guidance for Sociological Change.* Nashville. GlobalEd AdvancePRESS

Gutek, Gerald, (1988). *Philosophies, Ideologies, and Theories of Education*, Allyn & Bacon.

Hamby, B.W. (2007). *The Philosophy of Anything: Critical Thinking in Context.* Dubuque, Iowa. Kendall Hunt Publishing Company.

Hammond M, Collins R (1991). *Self-directed Learning: Critical Practice.* Kogan Page.

Israel, M. and Hay, I. (2006). *Research Ethics for Social Scientists: Between Ethical Conduct and Regulatory Compliance.* London: Sage.

Kenneth D. Bailey (2006). *Living Systems Theory and Social Entropy Theory.* Systems Research and Behavioral Science, 22, 291-300.

Leithwood, K. (Ed.) (2000). *Understanding Schools as Intelligent Systems.* CT: JAI Press

McLuhan, Marshall (1967).*The Medium is the Message.* London: Allen Lane

Miller, James Grier, (1978). *Living Systems.* New York: McGraw-Hill.

O'Toole, James.(1995). *Leading Change: Overcoming the Ideology of Comfort and The Tyranny of Custom.* San Francisco: Jossey-Bass Publishers.

Pohl, Michael. (1999). *Learning to Think, Thinking to Learn: Models and Strategies to Develop a Classroom Culture of Thinking.* Hawker Brownlow Ed.

Ramjattan, Subesh and Green, H.L. (2019) *God Has Confidence in You – No Test Has Come Your Way, But Such As Is Common To Man.* Nashville. GlobalEd AdvancePRESS

Ruane, Janet. M. (2004). *Essentials of Research Methods: A Guide to Social Science Research.* Blackwell.

Schaller, Lyle. (1972). *The Change Agent.* Nashville: Abingdon.

Seech, Zachary (2005), *Open Minds and Everyday Reasoning*, 2nd Edition. Belmont, CA: Wadsworth/ Thomson Learning.

Shook, John. (2000). *Dewey's Empirical Theory of Knowledge and Reality.* The Vanderbilt Library of American Philosophy.

Sleeper, R.W. (2001). *The Necessity of Pragmatism: John Dewey's Conception of Philosophy.* Introduction by Tom Burke. University of Illinois Press.

Swanson, G.A. and Miller, James Grier. (1989) *Measurement and Interpretation in Accounting: A Living Systems Theory Approach.* New York: Qurum Books.

Swanson, G.A., and Green, Hollis L. (1991, 2023). *Understanding Scientific Research: An Introductory Handbook for the Social Professions.* Nashville: Oxford/ACRSS Press.

White, Alasdair A. K. (2008). *From Comfort Zone to Performance Management.* White & MacLean Publishing.

Wlodkowski, R. J. (1998). *Enhancing Adult Motivation to Learn.* San Francisco, Jossey-Bass Publications.

ONLINE RESOURCES

Online International Journals Research methods knowledge database. http://www.socialresearchmethods.

Resource for methods in evaluation in social research. http://gsociology.icaap.org/methods/

Research methods and statistics arena. http://www.researchmethodsarena.com/resources/

Quantitative and Qualitative Analysis in Social Sciences. http://www.qass.org.uk/

Social Research Update. http://sru.soc.surrey.ac.uk/

Survey Research Methods. http://w4.ub.uni-konstanz.de/srm/

A Faith-based teacher asked –
"How does one get to heaven?"
A child answered –
"Turn right and go straight!"
"Correct," said the teacher –
"How do we walk on God's pathway?"
A child answered –
"Step by step claiming His Promises!"
"How long does it take?"
"Until the journey ends and
Heaven's Gate opens!"

GLOBAL EDUCATIONAL ADVANCE, INC.

GlobalEdAdvance.org

gea-books.com

GlobalEdAdvance@aol.com

www.ingramcontent.com/pod-product-compliance
Lightning Source LLC
Chambersburg PA
CBHW061757110426
42742CB00012BB/1912